Principles of
WATER QUALITY CONTROL

FIFTH EDITION

T. H. Y. TEBBUTT
PhD, SM, BSc, CEng, FICE, FCIWEM

Professor of Water Management,
School of Construction, Sheffield Hallam University
Past President,
Chartered Institution of Water and Environmental Management

BUTTERWORTH
HEINEMANN

Butterworth-Heinemann
Linacre House, Jordan Hill, Oxford OX2 8DP
225 Wildwood Avenue, Woburn, MA 01801-2041
A division of Reed Educational and Professional Publishing Ltd

℞ A member of the Reed Elsevier plc group

OXFORD   BOSTON   JOHANNESBURG
MELBOURNE   NEW DELHI   SINGAPORE

First published 1971
Reprinted 1975
Second edition 1977
Reprinted 1979
Third edition 1983
Reprinted 1991
Fourth edition 1992
Fifth edition 1998
Reprinted 1998

British Library Cataloguing in Publication Data
A catalogue record for this book is available from the British Library

Library of Congress Cataloguing in Publication Data
A catalogue record for this book is available from the Library of Congress

ISBN 0 7506 3658 0

Composition by Genesis Typesetting, Rochester, Kent
Printed and bound in Great Britain by Biddles Ltd, Guildford and King's Lynn

Contents

Preface to the fifth edition

Since the publication of the first edition of this book in 1971 there has been a remarkable growth in public interest in environmental matters. The general public is now accustomed to reading articles on environmental management in the press and programmes on radio and television often feature environmental topics. The advent of green politics has resulted in increased environmental legislation which has had significant impacts on water quality control. Sustainable development has become a common topic for discussion in the media and is strongly supported in many circles. Some of this public debate has been soundly based, but some has lacked scientific and technological credibility and the increasing costs of environmental control may show a poor return in terms of the quality of life. The growth of environmental legislation and regulation in many countries, sometimes coupled with transfer of public sector activities to the private sector, has further focused attention on environmental management. In such circumstances it is important that the fundamental factors controlling water quality are fully understood and this text continues to highlight the relationships between theory and application. In developed countries concerns have emerged about the presence of micropollutants in water supplies, but it must be remembered that in many parts of the world such concerns are irrelevant when the limited available water supplies frequently spread enteric diseases. Although this book is primarily concerned with developed country technology, the specialized aspects of developing country water supply and sanitation have been outlined where appropriate and a separate chapter covers simple water supply and sanitation systems.

The book is intended as a text for undergraduate courses in civil and environmental engineering, environmental management and environmental sciences and as preliminary reading for graduate students on environmentally based taught courses. It should also be useful for readers wishing to improve their understanding of water quality control for post-experience training and professional development courses such as the Diploma of the Chartered Institution of Water and Environmental Management. In this latest edition several chapters have largely been rewritten and others have been fully revised with additional worked examples where appropriate and updated recommendations for further reading. The original concept of providing a concise description of the science of water quality control has been retained and it is hoped that this latest edition will continue to achieve this objective.

T. H. Y. T.
School of Construction
Sheffield Hallam University

Preface to the first edition

This book is designed as a text for undergraduate civil engineering courses and as preliminary reading for postgraduate courses in public health engineering and water resources technology. It is hoped that it will also be of value to workers already in the field and to students preparing for the examinations of the Institute of Water Pollution Control and the Institution of Public Health Engineers. The text is based on my own lecture courses to undergraduate civil engineers augmented by material prepared for extramural short courses. Wherever possible, simple illustrations have been used to clarify the text. Reproduction of detailed working drawings has been deliberately avoided since in my experience these are often confusing to the student until the fundamentals of the subject are fully understood. Problems with answers have been included throughout the book so that readers can check their understanding of the text. The SI system of units has been adopted and it is hoped that the complete absence of Imperial units will encourage familiarity with their metric replacements.

The preparation of this book owes much to the enthusiasm for the subject which I gained from my first mentor, Professor P. C. G. Isaac of the University of Newcastle upon Tyne. I am most grateful to my colleague M. J. Hamlin for his helpful comments on the text and to Vanessa Green for her expert typing of the manuscript.

T. H. Y. T.
Department of Civil Engineering
University of Birmingham

1

Water – a precious natural resource

Water is the most important natural resource in the world since without it life cannot exist and most industries could not operate. Although human life can exist for many days without food, the absence of water for only a few days has fatal consequences. The presence of a safe and reliable source of water is thus an essential prerequisite for the establishment of a stable community. In the absence of such a source a nomadic life style becomes necessary and communities must move from one area to another as demands for water exceed its availability. It is therefore not surprising that sources of water are often jealously guarded and over the centuries many skirmishes have taken place over water rights. History shows many occasions where agricultural development has been hindered by interference with water supplies as part of the conflict between landowners and settlers which has occurred in numerous parts of the world. Other conflicts in relation to water supplies can arise because of the effects which human and industrial wastes can have on the environment. This means that the importance of water as a natural resource which requires careful management and conservation must be universally recognized. Although nature often has great ability to recover from environmental damage, the growing demands on water resources necessitate the professional application of fundamental knowledge about the water cycle to ensure the maintenance of quality and quantity.

1.1 The development of water and wastewater services

The importance of safe water supply and effective sanitation was recognized centuries ago by several ancient civilizations. Archaeological excavations in Asia and the Middle East have revealed highly developed communities with piped water supplies, latrines and sewers. The Minoan civilization in Crete, which flourished 4000 years ago, used fired clay water and wastewater pipes and the dwellings were provided with flushing toilets. The Romans were expert public health engineers and had highly developed water supply and drainage systems in the main cities. Considerable amounts of water were used in continuously operating fountains which provided the main water supply for most of the population, although wealthy families had their own piped supplies. To satisfy the demands for water, many urban areas in the Roman Empire benefited from the construction of major aqueducts, of which the 50 km long Nimes Aqueduct and the still surviving Pont du Gard are fine examples. The concept of transporting good quality water from upland catchments with a high rainfall into an urban area thus has a long history and demonstrates an important concept in water management. As well as water supply systems, Roman cities had stone sewers in the streets which collected both surface runoff and discharges from latrines for conveyance beyond the city limits. There is, however, little evidence that the Romans provided any treatment for the wastewater from their cities and thus their understanding of environmental protection was probably somewhat limited.

With the demise of the Roman Empire most of its public works facilities eventually fell into disuse and for centuries water supply and sanitation received scant attention from legislators and the general public. In Europe the Middle Ages often saw the establishment of towns at crossing points on rivers and these watercourses usually provided a convenient source of water and an apparently convenient repository for liquid and solid wastes. Although sewers were constructed, they were intended solely for the carriage of surfacewater and in the UK the discharge of foul sewage to the sewers was forbidden by law until 1815. Sanitary provisions in both rural and urban communities were minimal, records for London in 1579 show that one street with sixty houses had three communal latrines. Discharges of liquid and solid wastes from house windows were a common hazard to passers-by and it is not surprising that life expectancy for most people was not much more than thirty-five years. Lack of facilities in sparsely populated rural areas did not always cause major problems, but the rapid growth of urban populations, often in very poor accommodation, caused major hazards to public health.

By the middle of the last century sanitary conditions in most large towns and cities in Europe were appalling, with the rapid and sometimes catastrophic spread of water-related diseases. The general debilitating effects of these diseases, together with the consequences of poor nutrition and badly constructed housing, resulted in virtually continuous states of ill health in the urban populations, with children being particularly prone to succumb to an early death. Sir Edwin Chadwick was commissioned to investigate the situation and in his 1842 report he concluded that health depended upon sanitation, sanitation was an engineering matter requiring improved water supply to houses and a proper arterial drainage system, a single authority should administer all sanitary matters in an area, and expert advisers in engineering and medical matters were essential. Sir Edwin can thus be called the modern father of the disciplines of public health and public health engineering. In an attempt to improve matters, a law was passed in 1847 which made it obligatory in London for cesspit and latrine wastes to be discharged to the sewers. Unfortunately, most of London's sewers drained to the Thames from which some of the city's water supply was obtained. In addition, many of the sewers were badly constructed and maintained so that a good deal of their contents tended to leak out into the surrounding shallow aquifers which also provided water supplies. Little notice was taken of Chadwick's conclusions and thus the inevitable consequence was that urban watercourses and aquifers became increasingly contaminated with sewage. Rivers were objectionable to both sight and smell, but, more importantly, waterborne outbreaks of cholera became rampant in the cities of Europe with thousands of deaths every year. Similar situations arose in the growing cities of North America, and Lemel Shattuck in 1850 reported on public health problems in Massachusetts. He too saw the need for collaboration between the engineering and medical professions to achieve improvements.

In London the Broad Street Pump episode which contributed to the 10 000 cholera deaths in 1854 provided the evidence for Dr John Snow to demonstrate the link between sewage pollution of drinking water and the presence in the community of cholera. Although microorganisms had been observed by van Leeuwenhoek in 1680 with his microscope their true nature had not been understood. The existence of bacteria as living organisms and their role in disease was demonstrated by Pasteur in 1860 and in 1876 Koch developed culture techniques for the growth and identification of microbial species. By the 1860s it was realized that Chadwick's concepts of constant water supply and efficient sewerage systems could provide solutions to the growing health problems, although by then he had left his public offices. Public and parliamentary concerns resulted in the commissioning of the first major public health engineering works of modern times, Bazalgette's intercepting sewers which collected London's sewage for conveyance and discharge to the tidal reaches of the Thames and the transfer of water abstraction to points upstream of the tidal limit. Thus by 1870 waterborne outbreaks of disease had been greatly reduced in the UK and similar developments were taking place in other European countries and in the cities of North America. The Industrial Revolution further encouraged the growth of urban populations and hastened the need for major water supply schemes. Many of these relied on the Roman concepts of upland catchments and long aqueducts as exemplified by the Elan Valley scheme for Birmingham, Loch Katrine for Glasgow and the Croton and Catskill reservoirs for New York.

Only by continual and costly attention to water quality control has it been possible virtually to eradicate waterborne diseases from developed countries. A major result of this achievement has been that life expectancy in most European countries has almost doubled since 1850. Although advances in medical science have played some part in this improvement in life span, the role of environmental engineers and scientists in providing safe water supply and effective sanitation has been the major factor. Such successes should not, however, be allowed to mask the enormous tasks which remain to be tackled. A survey in 1975 found that 80 per cent of the world's rural population and 23 per cent of the urban population did not have reasonable access to a safe water supply. The sanitation situation was even worse with 85 per cent of the rural population and 25 per cent of the urban population having no provision for sanitation at all. The growth of populations in many developing countries was such that unless strenuous efforts to increase water supply and sanitation facilities could be made, the percentage of the world's population with satisfactory services would actually decrease in the years to come. The United Nations Organization therefore designated the period 1981–90 as the International Drinking Water Supply and Sanitation Decade with the aim of providing safe drinking water and adequate sanitation for all by 1990. Such an aim, although laudable, was unrealistic and the lack of sufficient trained personnel coupled with the worsening world economic situation meant that there has been relatively little improvement in the percentage of the world's population with an acceptable level of service. By the end of 1990 over 1300 million more people had clean water than in 1980 and over 750 million more people had improved sanitation provisions. Unfortunately, the birth rate in much of the developing world is such that the increased number of people served has not kept pace with the growth in population. The UN then launched a second programme called Safe Water 2000, which has more pragmatic objectives and which involves setting achievable targets. There will be more emphasis on cost sharing, relevant technology and sociological aspects of water supply and sanitation. There is still much to do before the apparently simple requirements of safe drinking water and adequate sanitation are available to all.

1.2 Sustainable development

In the developed world environmental matters now receive a great deal of public attention and the environment has taken on political implications. Although population growth is usually low in such countries, so that demands for water are not increasing greatly, there are a number of problems which focus attention on water quality control. Improved analytical techniques can now reveal the presence in water of hundreds of trace chemicals which arise from industrial processes and also as a consequence of some water and wastewater treatment processes. Increased leisure time places greater pressures on water-based recreation facilities and media interest in the environment stimulates public perception of water quality topics. Better understanding of food chains and the ability to understand complex biochemical and ecological reactions has resulted in water industries in many countries being subjected to more stringent restrictions on their operations and levels of service. The possibility of the greenhouse effect and changes in the ozone layer, producing far-reaching alterations in our environment, are further causes for public discussion and concern. In most developed countries there is an appreciation that environmental matters pose complex problems and there is a need for an overall view of the topic. In the less developed countries, although the need for environmental protection is recognized in some circles, the apparently more urgent pressures of population growth and economic survival usually control the situation. It has become clear that many of the environmental problems which arose in developed countries did so because of a lack of appreciation, concern and understanding of the causes of environmental pollution. International discussions aimed at preventing the earlier mistakes of the developed countries being replicated throughout the rest of the world resulted in the introduction of the concept of sustainable development. The Brundtland report, *Our Common Future*, published in 1987 defined sustainable development as

> development that meets the needs of the present without compromising the ability of future generations to meet their own needs.

In essence this definition implies

- recognition of the essential needs – particularly of the world's poor
- concern for the establishment of social equity between generations and within generations
- recognition of the limitations imposed by the capabilities of technology and social organizations on the ability of the environment to meet present and future demands.

In relation to water the concepts set out above can be interpreted as follows.

1. Water is a scarce resource which should be viewed as both a social and an economic resource.
2. Water should be managed by those who most use it, and all those who have an interest in its allocation should be involved in the decision making.
3. Water should be managed within a comprehensive framework, taking into account its impact on all aspects of social and economic development.

If these concepts can be incorporated into policy and practice, predictions of ever increasing environmental degradation in a world with decreasing resources could be replaced by an era of economic growth based on policies that sustain and expand natural environmental resources.

The European Commission defines the objectives of a sustainable water policy as

- provision of a secure supply of safe drinking water in sufficient quantity
- provision of water resources of sufficient quality and quantity to meet other economic requirements of industry and agriculture
- quality and quantity of water resources sufficient to protect and sustain the good ecological state and functioning of the aquatic environment
- management of water resources to prevent or reduce the adverse impact of floods and minimize the effects of droughts.

In 1992 the United Nations Conference on Environment and Development – the 'Earth Summit' – at Rio de Janeiro agreed the Rio Declaration which sets out the fundamental objectives of a programme of international agreements which respect the interests of all and protect the integrity of the global environmental and developmental system. The ways in which sustainable development can be achieved are set out in twenty-seven clauses which cover the special needs of developing countries, the elimination of unsustainable patterns of production and consumption, the importance of public participation, the value of environmental impact analyses, the need for effective environmental legislation and the adoption of the precautionary principle in matters affecting the quality of the environment. Agenda 21, agreed at Rio, is an action plan for the next century which takes into account the different needs of developed and developing countries. It gives high level political commitments to realistic, achievable and measurable goals for integrating environmental concerns into a broad range of activities including industry, agriculture, energy, fisheries, land use, water resources management and waste treatment. Agenda 21 includes a commitment to regular reporting by countries of their progress towards the agreed objectives.

Many of the matters which have been highlighted in the growing environmental discussion cannot be brought to a clear decision at our current state of knowledge. Informed discussion of matters affecting our environment is obviously to be encouraged but, unfortunately, there are occasions where media comments are misplaced or even mischievous. It is important that, where investments are to be made with the aim of improving public health or of protecting the environment, a realistic cost–benefit analysis is carried out to justify decisions. Although decisions may eventually be made on political or philosophical grounds, it is essential that the fullest engineering and scientific information is available to assist the decision makers.

1.3 Water resources

Water is a finite natural resource and in the context of a tripling of global water use since 1950 many parts of the world are facing growing pressures on their water resources. In Europe the demand for water has increased from $100\,km^3$ a year in 1950 to $550\,km^3$ a year in 1990 with a predicted rise to $650\,km^3$ a year by 2000. In such circumstances, over abstraction from surface and underground supplies may provide short-term solutions but they are not sustainable in the longer term. The science of hydrology is concerned with the assessment of water resources in the hydrological cycle (Figure 1.1) and their management for the optimum results. It will be appreciated that in any management plan for water resources it is vital to assess both the quality and quantity of the available supplies.

The earth and its surrounding atmosphere contain large amounts of water and about 7 per cent of the earth's mass is made up of water. However, 96.5 per cent of this water occurs as saline seawater and much of the remaining freshwater is incorporated into the polar icecaps and glaciers. Only about 0.7 per cent of the earth's water exists as freshwater in lakes, rivers, shallow aquifers and in the

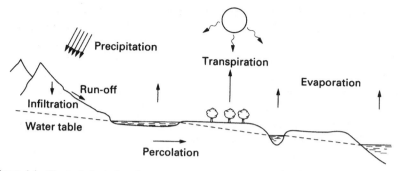

Figure 1.1 The hydrological cycle.

atmosphere. It is this water which takes part in the hydrological cycle and which fixes finite limits on availability. If this available water were distributed on the earth's surface in the same way as its population density there would be ample resources for all predicted needs. In practice, however, the spatial distribution of precipitation varies widely from many metres a year in mountainous tropical rain forests to essentially nil in major desert areas. This imbalance is well demonstrated by the fact that 20 per cent of the freshwater on the earth is found in the Amazon Basin which has only a minute percentage of the earth's population. Even within continents there are enormous variations between rainfall and population density. In general, heavy rainfall which produces high runoff and good groundwater recharge is found in mountainous regions with low population densities. Flat lowland areas which are favoured for both urban development and for agriculture are often in the rain shadow of mountains and thus usually have low precipitation. In the UK, for example, the Scottish Highlands has an average population density of about 2 persons per km^2 and precipitation can exceed 3 m a year. In South East England the population density exceeds 500 persons per km^2 but the rainfall is only about 0.6 m a year. Clearly in what is usually thought of as a wet country there can be very wide variations in water availability on a local or regional basis.

The renewable freshwater resources in the UK amount to about 2000 m^3 per person each year although for England and Wales the figure is only 1400 m^3 per person per year and for the Thames region it is only 250 m^3 per person per year. The concept of available freshwater is used by hydrologists and water resources planners to characterize the situation in a region and it is generally accepted that within the range 1000–2000 m^3 of freshwater per person a year there is stress on natural water resources. When availability falls below 1000 m^3 per person a year, water scarcity becomes evident with increasingly severe constraints on food production, economic development and environmental protection. Table 1.1 gives examples of water availability in a range of countries covering both water-rich and water-scarce situations. Direct consumption of water by people is actually a relatively small percentage of the total water demand. The use of water

Table 1.1 Some examples of water availability

Country	Freshwater resources $(10^3 m^3/person\ year)$
Water-rich	
Guyana	230
Liberia	90
Venezuela	44
Brazil	35
Ecuador	29
Burma	27
Cameroon	18
Guatemala	13
Nepal	10
Water-stressed	
Portugal	3.6
Ghana	3.4
Spain	2.8
Pakistan	2.7
India	2.3
UK	2.0
South Africa	1.4
Sudan	1.2
Germany	1.1
Water-scarce	
Belgium	0.8
Yemen	0.7
Algeria	0.7
Netherlands	0.6
Kenya	0.5
Israel	0.4
Singapore	0.2
Jordan	0.2
Saudi Arabia	0.1
Malta	0.08
Egypt	0.03
Bahrain	0

After Newson (1992), Overseas Development
Administration (1993) and Postel (1993).

for food production in agriculture is by far the most important global use and this use is particularly important in the developing countries. Agriculture consumes almost 65 per cent of all renewable water, industry around 20 per cent and public water supply only about 7 per cent.

1.4 The role of engineers and scientists

Public works such as water supply and sewage disposal schemes have traditionally been seen as civil engineering activities and water engineering is probably the largest single branch of the civil engineering profession. The connection with civil engineering is due to the fact that most water engineering works involve large structures and require a good understanding of hydraulics. Water science and technology is, however, an interdisciplinary subject involving the application of biological, chemical and physical principles in association with engineering techniques. Thus engineers and scientists who practise in water quality control must have a good appreciation of the interface between their individual disciplines and of the complex nature of many environmental reactions. The increasing amount of information which is required for the efficient design and operation of water quality control systems means that practitioners must also be conversant with developments in information technology. Solutions to environmental problems are rarely cost-free and thus the choice between various options must be made with an understanding of basic economic principles. Major water quality control projects are undertaken by a team of specialists from many disciplines who can bring their own particular expertise to the project whilst appreciating the need for collaborative work between disciplines to produce a cost-effective, environmentally acceptable solution.

A major objective in water quality control work is to reduce the incidence of water-related diseases. This objective depends on the ability to develop water sources to provide an ample supply of water of wholesome quality, i.e. a water free from

- visible suspended matter
- excessive colour, taste and odour
- objectionable dissolved matter
- aggressive constituents
- bacteria indicative of faecal pollution.

Drinking water supplies must obviously be fit for human consumption, i.e. of potable quality, and they should also be palatable, i.e. aesthetically attractive. In addition, as far as is feasible, public water supplies should be suitable for other domestic uses such as clothes washing, and so on.

Having provided a water of suitable quality and quantity. by source protection and development and by the application of appropriate treatment processes, it becomes necessary to convey the supply to consumers via a distribution system comprising water mains, pumping stations and service reservoirs. Most domestic and industrial uses of water cause a deterioration in quality with the resultant production of a wastewater which must be collected and given suitable treatment before release to the environment. In many situations treated wastewaters provide a significant proportion of the water resources for other users. Figure 1.2 illustrates in diagrammatic form typical water supply and wastewater disposal systems. The provision of water and wastewater services is a major process industry which in England and Wales

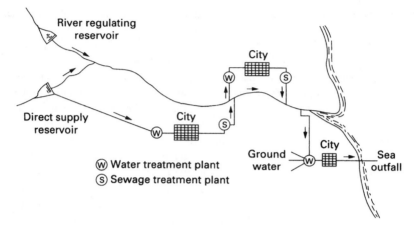

Figure 1.2 Water supply and wastewater disposal systems.

handles around 15 million m^3 of water and of wastewater each day at a total cost to customers of around £1.5 per m^3.

As will be discussed in Chapter 7, water has many uses so that any quality management or regulatory system has to consider numerous requirements and constraints. Water quality control measures must strike a balance between the needs of water supply services and effluent discharge requirements. Fisheries must be preserved and conservation of the water environment must be encouraged. The amenity aspects of bodies of water are becoming of increasing importance in developed countries as are recreational uses for a variety of sports and hobbies. All of these factors must be recognized in a situation where industrial activity, changing agricultural practices and increasing urbanization can have considerable influences on water quality.

Urban developments produce large volumes of solid wastes which can pose major environmental difficulties in their disposal. Landfills and other solid waste disposal sites can be responsible for major water pollution control problems since rainfall or groundwater can leach highly contaminating material from the deposited materials. Attention is increasingly being directed towards the reduction of excessive packaging materials and the recovery and re-use of other waste material. There is a need for what can be termed 'clean technology' to substitute 'cleaner' products and processes for those which contribute in a major way towards environmental pollution. It is important to realize that many environmental contaminants may affect air, land and water and care must thus be taken to ensure that a solution for pollution control in one phase does not produce problems elsewhere. Such problems may often be on an international scale because of the circulations in the atmosphere and in the oceans. The concept of integrated pollution control (IPC) is highly relevant to the effective conservation and management of the global environment.

References

Newson, M. (1992). *Land, Water and Development*, p. 235. London: Routledge.

Overseas Development Administration (1993). *A Fresh Approach to Water Resources Development*. London: ODA.

Postel, S. (1993). Facing water scarcity. In *State of the World 1993*, p. 22. New York: Norton and Company.

Further reading

Acheson, M. A. (1990). The Chadwick Centenary Lecture—A review of two centuries of public health. *J. Instn Wat. Envir. Managt*, **4**, 474.

Bailey, R. A. (1991). *An Introduction to River Management*. London: IWEM.

Bailey, R. A. (1997). *An Introduction to Sustainable Development*. London: CIWEM.

Barty-King, H. (1992). *Water – The Book*. London: Quiller Press.

Binnie, G. M. (1981). *Early Victorian Water Engineers*. London: Thomas Telford Ltd.

Binnie, G. M. (1987). *Early Dam Builders in Britain*. London: Thomas Telford Ltd.

Bramley, M. (1997). Future issues in environmental protection: A European perspective. *J. C. Instn Wat. Envir. Managt*, **11**, 79.

Clarke, K. F. (1994). Sustainability and the water and environmental manager. *J. Instn Wat. Envir. Managt*, **8**, 1.

Cook, J. (1989). *Dirty Water*. London: Unwin Hyman.

Department of the Environment (1992). *This Common Inheritance – The Second Year Report*. London: HMSO.

Department of the Environment (1994). *Sustainable Development – The UK Strategy*. London: HMSO.

Hall, C. (1989). *Running Water*. London: Robertson McCarta.

Hartley, Sir Harold (1955). The engineer's contribution to the conservation of natural resources. *Proc. Instn Civ. Engnrs*, **4**, 692.

Isaac, P. C. G. (1980). Roman public health engineering. *Proc. Instn Civ. Engnrs*, **68**(1), 215.

Kinnersley, D. (1988). *Troubled Waters: Rivers, Politics and Pollution*. London: Hilary Shipman.

Kinnersley, D. (1994). *Coming Clean*. London: Penguin.

Latham, B. (1995). *An Introduction to Water Supply in the UK*. London: CIWEM.

National Rivers Authority (1994). *Water – Nature's Precious Resource*. Bristol: NRA.

Nicholson, N. (1993). *An Introduction to Drinking Water Quality*. London: CIWEM.

Overseas Development Administration (1996). *Water for Life*. London: ODA.

Price, M. (1996). *Introducing Groundwater*, 2nd edn. London: Chapman and Hall.

Rodda, J. C. (1995). Guessing or assessing the World's water resources. *J. C. Instn Wat. Envir. Managt*, **9**, 360.

Shaw, Elizabeth, M. (1990). *Hydrology in Practice*, 2nd edn. London: Chapman and Hall.

Twort, A. C. (1990). *Binnie and Partners: A Short History to 1990*. Redhill: Binnie and Partners.

Wilson, E. M. (1990). *Engineering Hydrology*, 4th edn. London: Macmillan.

World Commission on Environment and Development (1987). *Our Common Future*. Oxford: Oxford University Press.

Wright, L. (1960). *Clean and Decent*. London: Routledge and Kegan Paul.

2

Characteristics of waters and wastewaters

The chemical formula for water, H_2O, is widely recognized, but unfortunately it is somewhat of a simplification since water has several properties which cannot be explained by such a simple structure. At low temperatures, particularly, water behaves as if its molecular form was H_6O_3 or H_8O_4 held together by hydrogen bridges. As the temperature approaches freezing, these structural linkages become more important than the thermal agitation which encourages a looser association of molecules. This interaction of the two molecular forces results in the effect that ice is less dense than water and the fact that the density of water increases as the temperature rises from 0°C to 4°C and then decreases with further increases in temperature because of the greater effect of thermal agitation at higher temperatures. Two consequences of this density effect are the bursting of pipes during freezing conditions and the thermal stratification of lakes. In the latter context, seasonal warming of a body of water results in the formation of a density barrier to mixing so that in deep lakes a large volume of water may be virtually stagnant and of poor quality. When the air temperature falls, the water on the surface of the lake cools and eventually reaches a density close to that of the lower level with the result that the stable surface layer eventually mixes with the lower layer. This overturn is usually brought about by wind-induced mixing and can give rise to serious water quality problems as the stagnant bottom water is mixed with the good quality water from the surface.

Because of its molecular structure and its electrical properties of a very high dielectric constant and a low conductivity, water is capable of dissolving many substances, so that the chemistry of natural water is very complex. All natural waters contain varying amounts of other materials in concentrations ranging from minute traces at the ng/l level of trace organics in rain water, to around 35 000 mg/l in seawater. Wastewaters usually contain most of the dissolved constituents of the water supply to the area with additional impurities arising from the waste-producing processes. Thus, the human metabolism releases about 6 g of chloride each day so that with a water consumption of 150 l/person day domestic sewage will contain at least 40 mg/l more chloride than the water supply to the area. A typical raw sewage contains around 1000 mg/l of solids in solution and suspension and is thus about 99.9 per cent pure water. Seawater at 35 000 mg/l of impurities is apparently much more contaminated than raw sewage. This anomaly highlights the fact that a simple measure of the total solids

Table 2.1   Important characteristics for various samples

Characteristic	River water	Drinking water	Raw sewage	Sewage effluent
pH	X	X	X	X
Temperature	X	X	X	
Colour	X	X		
Turbidity	X	X		
Taste		X		
Odour	X	X		
Total solids	X	X		
Settleable solids			X	
Suspended solids			X	X
Conductivity	X	X		
Radioactivity	X	X		
Alkalinity	X	X	X	X
Acidity	X	X	X	X
Hardness	X	X		
Dissolved oxygen (DO)	X	X		X
Biochemical oxygen demand (BOD)	X		X	X
Chemical oxygen demand (COD)	X		X	X
Total organic carbon (TOC)	X		X	X
Volatile organic carbon (VOC)	X	X		
Assimilable organic carbon (AOC)		X		
Organic nitrogen			X	X
Ammonia nitrogen	X		X	X
Nitrite nitrogen	X	X	X	X
Nitrate nitrogen	X	X	X	X
Chloride	X	X		
Phosphate	X		X	X
Synthetic detergent	X		X	X
Bacteriological counts	X	X		

content of a sample is insufficient to specify its character. The prospect of swimming in seawater is rather more attractive than the same activity in raw sewage! To gain a true understanding of the nature of a particular sample it is thus usually necessary to measure several different properties by undertaking analyses under the broad headings of physical, chemical and biological characteristics. The cost of analytical work can be considerable and thus not all characteristics would be investigated for a particular sample. Table 2.1 gives examples of the properties most likely to be used for various types of sample and the most important properties are discussed in the following sections.

2.1 Physical characteristics

Physical properties are in many cases relatively easy to measure and some may readily be observable by the layman.

1. *Temperature*. Basically important for its effect on other properties, e.g. speeding up of chemical reactions, reduction in solubility of gases, amplification of tastes and odours, etc.
2. *Taste and odour*. Due to dissolved impurities, often organic in nature, e.g. phenols and chlorophenols. They are subjective properties which are difficult to measure.
3. *Colour*. Even pure water is not colourless; it has a pale green-blue tint in large volumes. It is necessary to differentiate between true colour due to material in solution and apparent colour due to suspended matter. Natural yellow colour in water from upland catchments is due to organic acids which are not in any way harmful, being similar to tannic acid from tea. Nevertheless, many consumers object to a highly coloured water on aesthetic grounds and coloured waters may be unacceptable for certain industrial uses, e.g. production of high-grade art papers.
4. *Turbidity*. The presence of colloidal solids gives liquid a cloudy appearance which is aesthetically unattractive and may be harmful. Turbidity in water may be due to clay and silt particles, discharges of sewage or industrial wastes, or to the presence of large numbers of microorganisms.
5. *Solids*. These may be present in suspension and/or in solution and they may be divided into organic matter and inorganic matter. Total dissolved solids (TDS) are due to soluble materials whereas suspended solids (SS) are discrete particles which can be measured by filtering a sample through a fine paper. Settleable solids are those removed in a standard settling procedure using a 1 litre cylinder. They are determined from the difference between SS in the supernatant and the original SS in the sample.
6. *Electrical conductivity*. The conductivity of a solution depends on the quantity of dissolved salts present and for dilute solutions it is approximately proportional to the TDS content, given by

$$K = \frac{\text{conductivity (S/m)}}{\text{TDS (mg/l)}} \qquad (2.1)$$

Knowing the appropriate value of K for a particular water, the measurement of conductivity provides a rapid indication of TDS content.
7. *Radioactivity*. Measurements of gross β and γ activity are routine quality checks. Naturally occurring radon (an α emitter) can be a possible long-term health hazard with some groundwaters.

2.2 Chemical characteristics

Chemical characteristics tend to be more specific in nature than some of the physical parameters and are thus more immediately useful in assessing the properties of a sample. It is useful at this point to set out some basic chemical definitions.

- *Atomic weight* – weight (mass) of an atom of an element referred to a standard based on the carbon isotope ^{12}C. Also 'relative atomic mass'.
- *Molecular weight* – total atomic weight of all atoms in a molecule.
- *Molar solution* – solution containing the gram molecular weight (mole) of the substance in 1 litre, signified by M.
- *Valence* – property of an element measured by the number of atoms of hydrogen that one atom of the element can hold in combination or displace.
- *Equivalent weight* – the quantity of a substance which reacts with a given amount of a standard, given by

$$\text{equivalent weight} = \frac{\text{molecular weight}}{Z} \tag{2.2}$$

where, for acids, Z = the number of moles of H^+ obtainable from 1 mole of acid; for bases, Z = the number of moles of H^+ with which 1 mole of base will react. (A mole is the molecular weight in grams.)
- *Normal solution* – solution containing the gram equivalent weight of the substance in 1 litre, signified by N.

Some important chemical characteristics are described below.

pH

The *intensity* of acidity or alkalinity of a sample is measured on the pH scale which actually indicates the concentration of hydrogen ions present.

Water is only weakly ionized, as shown by the equilibrium

$$H_2O \rightleftharpoons H^+ + OH^-$$

Since only about 10^{-7} molar concentrations of H^+ and OH^- are present at equilibrium, $[H_2O]$ (i.e. the concentration of H_2O) may be taken as unity. Thus

$$[H^+][OH^-] = K = 1.01 \times 10^{-14} \text{ mole/l at } 25°C \tag{2.3}$$

Because this relationship must be satisfied for all dilute aqueous solutions the acidic or basic nature of the solution can be specified by one parameter – the concentration of hydrogen ions. This is conveniently expressed by the function pH, given by

$$pH = -\log_{10}[H^+] = \log_{10}\frac{1}{[H^+]} \tag{2.4}$$

resulting in a scale from 0 to 14 with 7 as neutrality, below 7 being acid and above 7 being alkaline.

Many chemical reactions are controlled by pH and biological activity is usually restricted to a fairly narrow pH range of 5–8. Highly acidic or highly alkaline waters are undesirable because of corrosion problems and possible difficulties in treatment.

Oxidation–reduction potential (ORP)

In any system undergoing oxidation there is a continual change in the ratio between the materials in the reduced form and those in the oxidized form. In such a situation the potential required to transfer electrons from the oxidant to the reductant is approximated by

$$ORP = E^0 - \frac{0.509}{z} \log_{10} \frac{[\text{products}]}{[\text{reactants}]} \qquad (2.5)$$

where E^0 = cell oxidation potential referred to H = 0 and z = number of electrons in the reaction.

Operational experience has established ORP values likely to be critical for various oxidation reactions. Aerobic reactions show ORP values of > +200 mV, anaerobic reactions occur below +50 mV.

Alkalinity

Due to the presence of bicarbonate, HCO_3^-, carbonate, CO_3^{2-}, or hydroxide OH^-. Most of the natural alkalinity in waters is due to HCO_3^- produced by the action of groundwater on limestone or chalk

$$\underset{\text{insoluble}}{CaCO_3} + H_2O + \underset{\substack{\text{from soil} \\ \text{organisms}}}{CO_2} \rightarrow \underset{\text{soluble}}{Ca(HCO_3)_2}$$

Alkalinity is useful in waters and wastes in that it provides buffering to resist changes in pH. It is normally divided into caustic alkalinity above pH 8.2 and total alkalinity above pH 4.5. Alkalinity can exist down to pH 4.5 because of the fact that HCO_3^- is not completely neutralized until this pH is reached. The amount of alkalinity present is expressed in terms of $CaCO_3$. (See also p. 31.)

Acidity

Most natural waters and domestic sewage are buffered by a CO_2:HCO_3 system. Carbonic acid, H_2CO_3, is not fully neutralized until pH 8.2 and will not depress the pH below 4.5. Thus CO_2 acidity is in the pH range 8.2 to 4.5; mineral acidity (usually due to industrial wastes) occurs below pH 4.5. Acidity is expressed in terms of $CaCO_3$.

Hardness

This is the property of a water which prevents lather formation with soap and produces scale in hot water systems. It is due mainly to the metallic ions Ca^{2+} and Mg^{2+} although Fe^{2+} and Sr^{2+} are also responsible. The metals are usually associated with HCO_3^-, SO_4^{2-}, Cl^- and NO_3^-. Hardness may actually have a health benefit (see also p. 61), but economic disadvantages of a hard water include increased soap consumption and higher fuel costs. Hardness is expressed in terms of $CaCO_3$ and is divided into two forms,

- carbonate hardness: due to metals associated with HCO_3^-
- non-carbonate hardness: due to metals associated with SO_4^{2-}, Cl^-, NO_3^-.

The non-carbonate hardness is obtained by substracting the alkalinity from the total hardness.

If high concentrations of sodium and potassium salts are present, the non-carbonate hardness value may be negative, since such salts could form alkalinity without producing hardness.

Dissolved oxygen (DO)

Oxygen is a most important element in water quality control. Its presence is essential to maintain the higher forms of biological life and the effect of a waste discharge on a river is largely determined by the oxygen balance of the system. Unfortunately oxygen is only slightly soluble in water as indicated below for water with no chloride content and at a standard barometric pressure of 1 atm (760 mmHg or 1.013 bar).

Temp. (°C)	0	10	20	30
DO (mg/l)	14.6	11.3	9.1	7.6

This solubility is affected by the presence of chlorides which reduce the saturation dissolved oxygen concentration by about 0.015 mg/l per 100 mg/l of chloride at low temperatures (5–10°C) and by about 0.008 mg/l of chloride at higher temperatures (20–30°C). A barometric pressure correction must be made, which is directly proportional to the ratio of the actual pressure to the standard 760 mmHg. The fall in barometric pressure above sea level is approximately 80 mmHg per 1000 m of elevation.

Clean surface waters are normally saturated with DO, but such DO can be removed rapidly by the oxygen demand of organic wastes. Game fish require at least 5 mg/l DO and coarse fish will not exist below about 2 mg/l DO. Oxygen-saturated waters have a pleasant taste and waters lacking DO have an insipid taste; drinking waters are thus aerated if necessary to ensure maximum DO. For boiler feed waters DO is undesirable because its presence increases the risk of corrosion.

Oxygen demand

Organic compounds are generally unstable and may be oxidized biologically or chemically to stable, relatively inert, end products such as CO_2, NO_3, H_2O. An indication of the organic content of a waste can be obtained by measuring the amount of oxygen required for its stabilization using

1. biochemical oxygen demand (BOD) – a measure of the oxygen required by microorganisms whilst breaking down organic matter;
2. chemical oxygen demand (COD) – chemical oxidation using boiling potassium dichromate and concentrated sulphuric acid.

The results obtained usually show COD > BOD in magnitude and the magnitude of the BOD:COD ratio increases as biological oxidation proceeds.

Organic matter may be determined directly as total organic carbon (TOC) by specialized combustion techniques or by using the UV absorption characteristics of the sample. In both cases commercial instruments are readily available but are relatively costly to buy and operate. Volatile organic carbon (VOC) and assimilable organic carbon (AOC) are specialized characteristics used in taste and odour control and in the control of biological growths in water distribution systems. Because of the importance of oxygen demand considerations, the subject is covered in detail in Chapter 6.

Nitrogen

This is an important element since biological reactions can only proceed in the presence of sufficient nitrogen. Nitrogen exists in four main forms in the water cycle

- organic nitrogen – nitrogen in the form of proteins, amino acids and urea
- ammonia (NH_3) nitrogen – nitrogen as ammonium salts, e.g. $(NH_4)_2CO_3$, or as free ammonia
- nitrite (NO_2) nitrogen – an intermediate oxidation stage not normally present in large amounts
- nitrate (NO_3) nitrogen – final oxidation product of nitrogen.

Oxidation of nitrogen compounds, termed nitrification, proceeds thus

$$\text{Organic nitrogen} + O_2 \rightarrow \text{Ammonia nitrogen} + O_2 \rightarrow$$
$$NO_2 \text{ nitrogen} + O_2 \rightarrow NO_3 \text{ nitrogen}$$

Reduction of nitrogen, termed denitrification, may reverse the process

$$NO_3^- \rightarrow NO_2^- \rightarrow \begin{array}{c} \nearrow NH_3 \\ \searrow N_2 \end{array}$$

The relative concentrations of the different forms of nitrogen give a useful indication of the nature and strength of the sample. Before the availability of bacteriological analysis the quality of waters was often assessed by considering the nitrogen content. A water containing high levels of organic and ammonia nitrogen with little nitrite and nitrate nitrogen would be considered unsafe because of recent pollution. On the other hand, a sample with no organic and ammonia nitrogen and some nitrate nitrogen would be considered relatively safe, as nitrification had occurred and thus pollution could not have been recent.

Chloride

Chlorides are salts of hydrochloric acid or metals combined directly with chlorine. They are responsible for brackish taste in water and are an indicator of sewage pollution because of the chloride content of urine. The threshold level for chloride taste is 250–500 mg/l, although up to 1500 mg/l is unlikely to be harmful to healthy consumers who are accustomed to that concentration.

Trace organics

Over 600 organic compounds have been detected in raw water sources and most of them are due to human activity or industrial operations. Substances which have been found include benzene, chlorophenols, oestrogens, pesticides, polynuclear aromatic hydrocarbons (PAH) and trihalomethanes (THM). They are normally present in very low concentrations, but there is some concern about possible health effects if such materials were consumed over a long time even at trace levels.

When dealing with industrial wastewaters or their effects on watercourses and aquatic life many other specialized chemical characteristics may be important, including heavy metals, cyanide, oils and greases.

2.3 Biological characteristics

Living organisms play major roles in many aspects of water quality control and thus the assessment of the biological characteristics of a water is often of great significance. Because of this significance the subject of aquatic microbiology and ecology is discussed in Chapter 4. It will suffice here to note that the bacteriological analysis of drinking water supplies usually provides the most sensitive quality assessment. Raw sewage contains millions of bacteria per millilitre and many organic wastewaters have large populations of bacteria, but the actual numbers are rarely determined. Conventional treatment methods for sewage and organic wastewaters rely on the ability of microorganisms to stabilize organic matter so that very large numbers of microorganisms are found in wastewater treatment plants and in their effluents. Microorganisms can thus play valuable roles in wastewater treatment and sometimes also in water treatment, but they are usually considered as sources of potential nuisance and hazard in relation to drinking water.

2.4 Typical characteristics

Since waters and wastewaters vary widely in their characters it is not desirable to give specifications for what might be termed 'normal' samples. It is perhaps useful, however, to give some examples of typical water and wastewater qualities. Table 2.2 gives an indication of the characteristics which would be expected from three common sources of water and Table 2.3 outlines the characteristics of a typical sewage at various stages of treatment. Figure 2.1 gives a diagrammatic representation of the nature of domestic sewage.

Table 2.2 Typical characteristics of various water sources

Characteristic	Source		
	Upland catchment	Lowland river	Chalk aquifer
pH	6.0	7.5	7.2
Total solids (mg/l)	50	400	300
Conductivity (μS/cm)	45	700	600
Chloride (mg/l)	10	50	25
Alkalinity, total (mg/l HCO_3)	20	175	110
Hardness, total (mg/l Ca)	10	200	200
Colour (°H)	70	40	<5
Turbidity (NTU)	5	50	<5
Ammonia nitrogen (mg/l)	0.05	0.5	0.05
Nitrate nitrogen (mg/l)	0.1	2.0	0.5
DO (% saturation)	100	75	2
BOD (mg/l)	2	4	2
Colonies/ml at 22°C	100	30 000	10
Colonies/ml at 37°C	10	5 000	5
Coliform organisms/100 ml	20	20 000	5

°H = Hazen colour units; NTU = nephelometric turbidity units.

Table 2.3 Typical sewage analyses

Characteristic (mg/l)	Source		
	Crude	Settled	Final effluent
BOD	300	175	20
COD	700	400	90
TOC	200	90	30
SS	400	200	30
Ammonia nitrogen	40	40	5
Nitrate nitrogen	<1	<1	20

Another way of appreciating the significance of water quality parameters is to consider the various standards and guidelines which are used to specify water qualities for various uses. In the case of potable supply it is accepted practice to use guidelines or standards which are based on an assessment of the importance of a particular parameter or group of parameters. In this context it is useful to classify constituents of water into five groups.

1. *Organoleptic parameters* – characterized by being readily observable by the consumer but usually having little health significance; typical examples are colour, suspended matter, tastes and odours. Guidelines are normally set on the basis of aesthetic considerations.
2. *Natural physico-chemical parameters* – normal characteristics of waters such as pH, conductivity, dissolved solids, alkalinity, hardness, dissolved oxygen, etc. Some of these parameters may have limited health significance but in general the guidelines are intended to ensure a chemically balanced water.

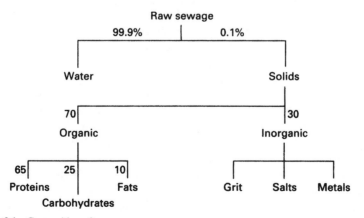

Figure 2.1 Composition of sewage.

3. *Substances undesirable in excessive amounts* – this group includes a wide variety of substances, some of which may be directly harmful in high concentrations, others may produce undesirable tastes and odours and some may not be directly troublesome in themselves but are indicators of pollution. Constituents in this group include chloride, fluoride, iron, manganese, nitrate, phenol and total organic carbon. Allowable levels for these substances are based either on consumer acceptability or on their significance in relation to other factors.

4. *Toxic substances* – a considerable number of inorganic and organic chemicals can have toxic effects on consumers of water containing them, the severity of the effects depending for a particular substance on the dose received, the period of consumption, and other dietary and environmental factors. Since the main concern in drinking water is with the long-term effects of exposure to low levels of potentially toxic materials it is not easy to set limits on a scientific basis. It is thus common for large factors of safety to be employed. Constituents which may be considered toxic include arsenic, cyanide, lead, mercury, organophosphorus compounds, pesticides, trihalomethanes.

5. *Microbiological parameters* – in most parts of the world these parameters are by far the most important in determining the safety of drinking water. Microbiological standards for drinking water are based on the need to ensure the absence of bacteria indicative of pollution by human wastes.

Table 2.4 gives examples of European Community standards for surface-waters used as raw water for drinking water. Table 2.5 shows the UK drinking water supply regulations which put into force the requirements of the European Community directive 80/778/EEC on water for drinking which are currently under review. For parameters where the EC directive gives only guide levels the UK regulations set prescribed concentrations and they require strict compliance with all EC maximum allowable concentrations (MACs) unless more stringent prescribed concentrations are specified. In the United States the 1974 Safe Drinking Water Act required the establishment of national drinking water quality regulations by the US Environmental Protection Agency. These regulations set standards and monitoring requirements for a wide range of health-related parameters which are being phased in over a number of years. In most cases the levels adopted are similar to those used by the EC and by the World Health Organization. The US regulations also require mandatory disinfection and filtration unless in the latter case it can be demonstrated to be unnecessary. Table 2.6 gives examples of World Health Organization guidelines for health-related parameters in drinking water quality. Table 2.7 gives details of parameters identified by WHO as likely to cause consumer complaints although they are not specifically health-related. In the WHO guidelines the term 'action level' is used to indicate a level above which the reasons for the presence of the substance should be investigated and remedial measures applied as appropriate. Other uses of water may also be subject to guidelines

Table 2.4 Surfacewater quality for drinking water abstraction (directive 76/464/EEC)

Treatment type[†]	A1		A2		A3	
Parameter (mg/l unless noted)	GL	MAC	GL	MAC	GL	MAC
pH units	6.5–8.5		5.5–9.0		5.5–9.0	
Colour units	10	20	50	100	50	200
SS	25					
Temp (°C)	22	25	22	25	22	25
Conductivity (µS/cm)	1000		1000		1000	
Odour (TON)	3		10		20	
Nitrate (NO$_3$)	25	50		50		50
Fluoride	0.7–1.0	1.5	0.7–1.7		0.7–1.7	
Iron (soluble)	0.1	0.3	1.0	2.0	1.0	
Manganese	0.05		0.1		1.0	
Copper	0.02	0.05	0.05		1.0	
Zinc	0.5	3.0	1.0	5.0	1.0	5.0
Boron	1.0		1.0		1.0	
Arsenic	0.01	0.05		0.05	0.05	0.1
Cadmium	0.001	0.005	0.001	0.005		0.05
Chromium (total)		0.05		0.05		0.05
Lead		0.05		0.05		0.05
Selenium	0.01		0.01		0.01	
Mercury	0.0005	0.001	0.0005	0.001	0.0005	0.001
Barium		0.1		1.0		1.0
Cyanide		0.05		0.05		0.05
Sulphate	150	250	150	250	150	250
Chloride	200		200		200	
MBAS	0.2		0.2		0.5	
Phosphate (P$_2$O$_5$)	0.4		0.7		0.7	
Phenol		0.001	0.001	0.005	0.01	0.1
Hydrocarbons		0.05		0.2	0.5	1.0
PAH		0.0002		0.0002		0.001
Pesticides		0.001		0.0025		0.005
COD					30	
BOD (ATU)	<3		<5		<7	
DO (% satn)	>70		>50		>30	
Kjeldahl nitrogen	1		2		3	
Ammonium (NH$_4$)	0.05		1	1.5	2	4
Total coliforms/100 ml	50		5000		50 000	
Faecal coliforms/100 ml	20		2000		20 000	
Faecal streptococci/100 ml	20		1000		10 000	
Salmonella	Absent in 5 litres		Absent in 1 litre			

[†]A1: Simple physical treatment and disinfection, A2: Normal full physical and chemical treatment with disinfection, A3: Intensive physical and chemical treatment with disinfection. MAC compliance is at 95 percentile level with the non-compliant 5 percentile not exceeding 150 per cent of the MAC level. GL, guide level; MAC, maximum allowable concentration.
ATU = allylthiourea; MBAS = methylene blue active substance; TON = threshold odour number.

Table 2.5 UK Drinking Water Supply Regulations (1989)

Parameter	Maximum value
A	
Colour	20 mg/l Pt/Co scale
Turbidity	4 FTU
Odour	3 at 25°C Dilution number
Taste	3 at 25°C Dilution number
Temperature	25°C
Hydrogen ion	9.5–5.5 (min.) pH units
Sulphate	250 mg/l
Magnesium	50 mg/l
Sodium	150 mg/l
Potassium	12 mg/l
Dry residue	1500 mg/l at 180°C
Nitrate	50 mg/l as NO_3
Nitrite	0.1 mg/l as NO_2
Ammonium	0.5 mg/l as NH_4
Kjeldahl nitrogen	1 mg/l
Permanganate value	5 mg/l O_2
Total organic carbon	No significant increase
Dissolved/emulsified hydrocarbons	10 µg/l
Phenols	0.5 µg/l
Surfactants	200 µg/l
Aluminium	200 µg/l
Iron	200 µg/l
Manganese	50 µg/l
Copper	3000 µg/l
Zinc	5000 µg/l
Phosphorus	2200 µg/l
Fluoride	1500 µg/l
Silver	10 µg/l
B	
Arsenic	50 µg/l
Cadmium	5 µg/l
Cyanide	50 µg/l
Chromium	50 µg/l
Mercury	1 µg/l
Nickel	50 µg/l
Lead	50 µg/l
Antimony	10 µg/l
Selenium	10 µg/l
Pesticides	
individual	0.1 µg/l
total	0.5 µg/l
PAH	0.2 µg/l
C	
Total coliforms	0 number/100 ml
Faecal coliforms	0 number/100 ml
Faecal streptococci	0 number/100 ml
Sulphite-reducing clostridia	<1 number/20 ml
Colony counts	No significant increase in number/ml at 22 or 37°C
D	
Conductivity	1500 µS/cm at 20°C
Chloride	400 mg/l
Calcium	250 mg/l
Substances extractable in chloroform	1 mg/l
Boron	2000 µg/l
Barium	1000 µg/l
Benzo 3,4 pyrene	10 µg/l
Tetrachloromethane	3 µg/l
Trichloroethene	30 µg/l
Tetrachloroethene	10 µg/l
E	*Minimum value*
Total hardness	60 mg/l as Ca
Alkalinity	30 mg/l as HCO_3

Samples should not contain any concentrations above the values shown in **A** to **C** and the 12 month average concentrations should not exceed the values shown in **D**.
FTU = formazin turbidity units.

Table 2.6 WHO Drinking Water Quality Guidelines (1993)

Parameter	Guide level (mg/l unless shown)	Note
Microbiological		
Total coliforms/100 ml	0	95% absent over 12 month period
E. coli	0	
Inorganics		
Antimony	0.005	
Arsenic	0.01	
Barium	0.7	
Boron	0.3	
Cadmium	0.003	
Chromium	0.05	
Copper	2	
Cyanide	0.07	
Fluoride	1.5	Depends on local conditions
Lead	0.01	
Manganese	0.5	
Mercury	0.001	
Molybdenum	0.07	
Nickel	0.02	
Nitrate	50	Sum of ratio of concentration to GL should
Nitrite	3	not exceed 1 for both together
Selenium	0.01	
Organics (part list only)		
Carbon tetrachloride	0.002	
Dichloromethane	0.020	
Trichloroethene	0.040	
Benzene	0.010	
Toluene	0.700	
Ethylbenzene	0.300	
Acrylamide	0.0005	
Nitrilotriacetic acid	0.200	
Tributyltin oxide	0.002	
Pesticides (part list only)		
Atrazine	0.002	
DDT	0.002	
2,4-D	0.030	
Heptachlor	0.00003	
Pentachlorophenol	0.009	
Permethrin	0.020	
Simazine	0.002	
Mecoprop	0.010	
Disinfectants and disinfectant by-products (part list only)		
Monochloramine	3	
Di- and trichloramine	5	
Bromate	0.025	
Chlorite	0.200	
Bromoform	0.100	
Chloroform	0.200	
Chloral hydrate	0.010	
Cyanogen chloride	0.070	
Radioactive constituents		
Gross alpha activity (Bq/l)	0.1	
Gross beta activity (Bq/l)	1	

Table 2.7 Substances and parameters in drinking water that may give rise to complaints from consumers (WHO, 1993)

Parameter	Level for complaint (mg/l unless shown)	Reason
Physical		
Colour (TCU)	15	Appearance
Turbidity (NTU)	5	Appearance/effective disinfection
Taste and odour	–	Acceptable
Temperature	–	Acceptable
Inorganics		
Aluminium	0.2	Deposition/discoloration
Ammonia	1.5	Odour/taste
Chloride	250	Taste/corrosion
Copper	1	Staining
Hydrogen sulphide	0.05	Odour/taste
Iron	0.3	Staining
Manganese	0.1	Staining
Sodium	200	Taste
Sulphate	250	Taste/corrosion
Total dissolved solids	1000	Taste
Zinc	3	Appearance/taste
Organics		
Toluene	0.024–0.170	Taste/odour (HRGL 0.700)
Xylene	0.020–1.800	Taste/odour (HRGL 0.500)
Disinfectants and disinfectant by-products		
Chlorine	0.6–1.0	Taste/odour
2-chlorophenol	0.0001–0.010	Taste/odour

HRGL = health-related guide level, NTU = nephelometric turbidity units, TCU = true colour units.

Table 2.8 Bathing water quality standards (directive 76/110/EEC)

Parameter (mg/l except as noted)	Guide limit (90th percentile)	Mandatory limit (95th percentile)
Cadmium	0.0025	0.0025
Mercury	0.0003	0.0003
Dissolved oxygen	80–120 per cent saturation	
Faecal coliforms (MPN/100 ml)	100	2000
Total coliforms (MPN/100 ml)	500	10000
Salmonella (MPN/litre)		0
Faecal streptococci (MPN/100 ml)	100	
Enteroviruses (MPN/10 litres)		0

MPN = most probable number.

Table 2.9 Freshwater fish water quality standards (directive 78/659/EEC)

Parameter	Annual average dissolved concentration (mg/l except as noted)	
	Salmonid fish	Coarse fish
Arsenic	0.05	0.05
Cadmium	0.005	0.005
Chromium	0.005–0.05	0.15–0.25
Copper	0.001–0.028	0.001–0.028
Lead	0.004–0.02	0.05–0.25
Mercury	0.001	0.001
Nickel	0.05–0.2	0.05–0.2
Zinc	0.01–0.125	0.075–0.5
Phosphate	65	131
Ammonia (total)	0.031	0.16
Ammonia (free)	0.004	0.004
Nitrite	0.003	0.003
BOD	3	6
Residual chlorine	0.0068	0.0068
pH (units)	6–9	6–9
Temperature (°C)	21.5	28
Suspended solids	25	25

or standards and Table 2.8 shows some EC directive values for quality parameters in relation to bathing waters. Table 2.9 gives some examples of EC directive values for various parameters in relation to fish life.

Further reading

Department of the Environment (1989). *The Water Supply (Water Quality) Regulations 1989*. HMSO.

Gardiner, J. and Mance, G. (1984). *United Kingdom Water Quality Standards Arising from European Community Directives*, Technical Report 204. Medmenham: Water Research Centre.

Gardiner, T. and Zabei, T. (1989). *United Kingdom Water Quality Standards Arising from European Community Directives—An Update*, Report FR0041. Marlow: Foundation for Water Research.

Hammerton, D. (1996). *An Introduction to Water Quality in Rivers, Coastal Waters and Estuaries*. London: CIWEM.

Holden, W. S. (ed.) (1970). *Water Treatment and Examination*. London: Churchill.

Nicholson, N. (1993). *An Introduction to Drinking Water Quality*. London: CIWEM.

Sawyer, C. N. and McCarty, P. L. (1978). *Environmental Chemistry for Engineers*, 3rd edn. New York: McGraw-Hill.

Tebbutt, T. H. Y. (1983). *Relationship Between Natural Water Quality and Health*, Technical Documents in Hydrology. Paris: UNESCO.

World Health Organization (1993). *Guidelines for Drinking Water Quality*, 1, *Recommendations*, 2nd edn. Geneva: WHO.

3

Sampling and analysis

To obtain an accurate representation of the composition and nature of a water or wastewater it is first essential to ensure that the sample analysed is truly representative of the source. Having satisfied this requirement, it is then necessary to carry out the appropriate analytical procedures using standard techniques which have been developed specifically for water and wastewater analyses.

3.1 Sampling

The collection of a representative sample from a source of uniform quality poses few problems and a single grab sample will be satisfactory. A grab sample will also be sufficient if the purpose of sampling is simply to provide a spot check to see whether particular limits have been complied with. However, most raw waters and wastewaters are highly variable in both quality and quantity so that a grab sample is unlikely to provide a meaningful picture of the nature of the source. This point is illustrated in Figure 3.1 which shows typical flow and strength variations in a sewer. To obtain an accurate assessment in this situation it is necessary to produce

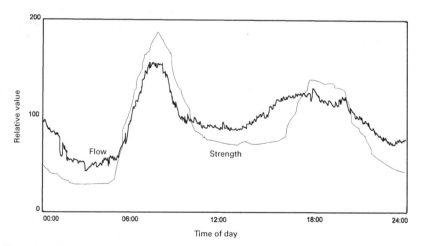

Figure 3.1 Typical flow and strength variations in a sewer during dry weather.

a composite sample by collecting individual samples at known time intervals throughout the period and measuring the flow at the same time. By bulking the individual samples in proportion to the appropriate flows an integrated composite sample is obtained. Similar procedures are often necessary when sampling streams and rivers and with large channel sections it may be desirable to sample at several points across the section and at several depths. Various automatic devices are available to collect composite samples and these may operate on either a time basis or on a flow-proportional basis. Sampling of industrial wastewater discharges may be even more difficult since they are often intermittent in nature. In these circumstances it is important that the nature of the operations producing the discharge is fully understood so that an appropriate sampling programme can be drawn up to obtain a true picture of the discharge.

When designing a sampling programme it is vital that the objective of the exercise be clearly specified, e.g. to estimate maximum or mean concentrations, to detect changes or trends, to estimate percentiles or to provide a basis for industrial effluent charges. The degree of uncertainty which can be tolerated in the answer must be specified and it is also necessary to bear in mind the resources available for sampling and analysis. For example, it may be found that reducing the uncertainty of the results by a few per cent might require twice the number of samples thus making the whole exercise uneconomic. It is therefore important to set a realistic level for the uncertainty of the results, based on the intended use. Ideally, all analyses should be carried out on the sample immediately after collection and certainly the quicker the analysis can be done the more likely it is that the results will be a true assessment of the actual nature of the liquid in situ. With characteristics which are likely to be unstable such as dissolved gases, oxidizable or reducible constituents, etc., the analyses must be carried out in the field or the sample must be suitably treated to fix the concentrations of unstable materials. Changes in the composition of a sample with time can be retarded by storage at low temperature (4°C) and the exclusion of light is also advisable. The more polluted a sample, the shorter the time which can be allowed between sampling and analysis if significant errors are to be avoided.

3.2 Analytical methods

Because of its solvent properties water could contain one or many of the thousands of inorganic and organic substances found in the environment. Some of these substances are naturally occurring, but many are the result of human activities. Most are, however, usually present only in small concentrations. Thus, although most determinations are based on relatively straightforward analytical techniques these often need to be adapted to the specialized needs associated with the quantitative analysis of low concentrations of impurities. In certain cases, rather than determining individual impurities it may be more convenient to utilize 'blanket' determinations which measure a whole range of similar substances. A typical example of this approach is in the determination of organic substances in a wastewater, for which it is customary simply to measure the amount of oxygen required to oxidize these substances. The oxygen consumption then provides a measure of the strength or polluting potential of the wastewater.

Physical and chemical analyses are carried out using a range of gravimetric, volumetric and colorimetric techniques or by using sensing electrodes and specialized instrumental methods. There is increasing use of automated techniques for laboratories with large throughputs of samples and remote monitoring for important quality parameters can be an important weapon in the battle to fight water pollution. Because of the need to determine the presence of some parameters at minute concentrations there is a growing use of sophisticated analytical instruments, particularly when trace organic substances are being determined. The analysis of samples for microbiological parameters requires the use of specialized procedures and these are described in Chapter 4.

When considering analytical methods it is important to appreciate the difference between accuracy and precision – accuracy measures the closeness of results to the true value whereas precision measures the reproducibility of results when the same sample is analysed repeatedly. It is quite possible to have precise methods which are inaccurate or methods with poor precision but which provide a reasonably accurate result with repeated analyses of the same sample. Analytical methods and instrument manufacturers normally indicate accuracy and precision values although these are often obtained under ideal conditions and may not always be achieved under routine operation.

Gravimetric analysis

This form of analysis depends upon weighing solids obtained from the sample by evaporation, filtration or precipitation. Because of the small weights involved, a balance accurate to 0.0001 g is required together with a drying oven to remove all moisture from the sample. Gravimetric analysis is thus not suited for field testing. Its main uses are for the measurement of

- *total and volatile solids*: a known volume of sample in a preweighed nickel dish is evaporated to dryness on a water bath, dried at 103°C (for wastewaters) or 180°C (for potable waters) and weighed. The increase in weight is due to the total solids. The loss in weight on firing at 500°C represents the volatile solids;
- *suspended solids* (SS): a known volume of sample is filtered under vacuum through a preweighed glass-fibre paper (Whatman GF/C) with a pore size of 0.45 or 1.2 μm. Total SS are given by the increases in weight after drying at 103°C and volatile SS (VSS) are those lost on firing at 500°C;
- *sulphate*: for concentrations above 10 mg/l it is possible to determine sulphate by precipitating barium sulphate after the addition of barium chloride. The precipitate is filtered out of the sample, dried and weighed.

Volumetric analysis

Many determinations in water quality control can be carried out rapidly and accurately by volumetric analysis, a technique which depends on the measurement of volumes of liquid reagent of known strength. The requirements for volumetric analysis are relatively simple and are listed below.

- A pipette to transfer a known volume of the sample to a conical flask.
- A standard solution of the appropriate reagent. It is often convenient to make the strength of the standard solution such that 1 ml of the solution is chemically equivalent to 1 mg of the substance under analysis.
- An indicator to show when the end point of the reaction has been reached. Various types of indicator are available, e.g. electrometric, acid–base, precipitation, adsorption and oxidation–reduction.
- A graduated burette for accurate measurement of the volume of standard solution necessary to reach the end point.

An example of the use of volumetric analysis is found in the determination of alkalinity and acidity. Only with strong acids and strong bases does neutralization occur at pH 7; with all other combinations the neutralization points occur at pH 8.2 and pH 4.5. The indicators normally adopted for acidity and alkalinity are phenolphthalein (pink above pH 8.2, colourless below pH 8.2) and screened methyl orange (green above pH 4.5 and purple below pH 4.5). For the most accurate determinations in acid–base titrations a pH meter may be used for direct indication of the end point. Using 0.02 N standard solutions for alkalinity, acidity and also hardness determinations, 1 ml of titrant solution contains 1 mg $CaCO_3$, the common denominator in which all these parameters are conventionally expressed.

Volumetric analysis can be useful in establishing the particular forms of alkalinity present in a sample. Neutralization of OH^- is complete at pH 8.2, whereas neutralization of CO_3^{2-} is only half completed at pH 8.2 and not fully completed until pH 4.5 is reached

$$CO_3^{2-} + H^+ \rightarrow HCO_3^- + H^+ \rightarrow H_2CO_3$$

Examination of the titration results will indicate the composition of the alkalinity present if it is assumed that HCO_3^- and OH^- cannot exist together. This is not strictly true and for accurate work more detailed considerations must be made. Reference to Figure 3.2 shows that the following possibilities exist.

1. OH^- alone will give an initial pH of about 10 and in this case,

   OH^- alkalinity = caustic alkalinity = total alkalinity.

2. CO_3^{2-} will give an initial pH of about 9.5, so

$$CO_3^{2-} \text{ alkalinity } = 2 \times \text{ caustic alkalinity } = \text{ total alkalinity.}$$

3. OH^- and CO_3^{2-} together will give an initial pH about 10, so

$$CO_3^{2-} \text{ alkalinity } = 2 \times \text{ titration from pH 8.2 to pH 4.5, and}$$
$$OH^- \text{ alkalinity } = \text{ total alkalinity } - CO_3^{2-} \text{ alkalinity.}$$

4. CO_3^{2-} and HCO_3^- together will give an initial pH > 8.2 and < 10.5, so

$$CO_3^{2-} \text{ alkalinity } = 2 \times \text{ caustic alkalinity and}$$
$$HCO_3^- \text{ alkalinity } = \text{ total alkalinity } - CO_3^- \text{ alkalinity.}$$

5. HCO_3^- alone will give an initial pH < 8.2, so

$$HCO_3^- \text{ alkalinity } = \text{ total alkalinity.}$$

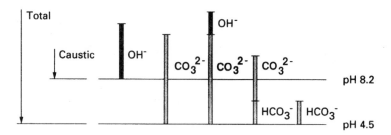

Figure 3.2   Forms of alkalinity.

Worked example for alkalinity determination

In an alkalinity determination using 0.02 N acid (1 ml = 1 mg calcium carbonate) a 100 ml sample of water gave the following results.

Titration to pH 8.2	14.5 ml acid
Total titration to pH 4.5	22.5 ml acid

Caustic alkalinity = $14.5 \times 1000/100 = 145$ mg/l
Total alkalinity   = $22.5 \times 1000/100 = 225$ mg/l

Titration to pH 8.2 = $2 \times$ titration from pH 8.2 to pH 4.5
        (14.5)   (22.5 − 14.5)

Hence from Figure 3.2 the alkalinity must be due to OH^- and CO_3^{2-}

CO_3^{2-} = 2 × (225 − 145) = 160 mg/l;
OH^- = 145 − (160/2) = 65 mg/l.

Other common uses of volumetric analysis are in the determination of chloride (silver nitrate with potassium chromate precipitation indicator), the Winkler dissolved oxygen determination (sodium thiosulphate with starch adsorption indicator) and in the COD determination (ferrous ammonium sulphate with Ferroin ORP indicator).

Colorimetric analysis

When dealing with low concentrations, colorimetric analyses are often particularly appropriate and there are many determinations in water quality control which can be quickly and easily carried out by this form of analysis.

To be of quantitative use a colorimetric method must be based on the formation of a completely soluble product with a stable colour. The coloured solution must conform with the following relationships.

1. Beer's law: Light absorption increases exponentially with the concentration of the absorbing solution.
2. Lambert's law: Light absorption increases exponentially with the length of the light path.

These laws apply to all homogeneous solutions and can be combined as

$$OD = \frac{I_0}{I} = abc \qquad (3.1)$$

where OD = optical density,
I_0 = intensity of light entering sample,
I = intensity of light leaving sample,
a = constant characteristic of particular solution,
b = length of light path in solution and
c = concentration of absorbing substance in solution.

The colour produced may be measured by a variety of methods.

Visual methods

1. Comparison tubes (Nessler tubes). A standard range of concentrations of the substance under analysis is prepared and the appropriate reagent added. The unknown sample is treated in the same manner and matched to the

standards by looking down through the solutions on to a white base. The procedure is time consuming since the standards fade and must be remade at intervals.
2. Colour discs. In this case the standards are in the form of a series of suitably coloured glass filters through which a standard depth of distilled water or sample without colour-forming reagents is viewed. The sample in a similar tube is compared with the colour disc and the best visual match selected.

Both of these methods are dependent upon somewhat subjective judgements so that reproducibility between different analysts may not be good. The colour disc method is very convenient for field use and a wide range of discs and prepacked reagents is available.

Instrumental methods

1. Absorptiometer or colorimeter. This type of instrument comprises a glass sample cell through which a beam of light from a low-voltage lamp is passed. Light emerging from the sample is detected by a photoelectric cell whose output is displayed on a meter. The sensitivity is enhanced by inserting in the light path a colour filter complementary to the solution colour and the range of measurement can be extended by using sample cells of different length.
2. Spectrophotometer. This is a more accurate type of instrument using the same basic principle as an absorptiometer but with a prism being employed to give monochromatic light of the desired wavelength. The sensitivity is thus increased and on the more expensive instruments measurements can be undertaken in the infrared and ultraviolet regions as well as in the visible light wavebands.

With both types of instrument a blank of the sample without the last colour-forming reagent is used to set the zero optical density position. The treated sample is then placed in the light path and the optical density noted. A calibration curve must be obtained by determining the optical density of a series of known standards at the optimum wavelength, obtained from analytical reference books or by experiments. In any form of colorimetric analysis it is important to ensure that full colour development has taken place before measurements are made and that any suspended matter in the sample has been removed. Suspended matter will of course prevent the transmission of light through a sample so that its presence will reduce the sensitivity of the determination and lead to erroneous results unless the blank has the identical concentration of suspended solids.

Turbidity in samples is usually determined by nephelometry, a photoelectric technique which measures the scattering of light by colloidal particles.

Electrode techniques

The measurement of such parameters as pH and oxidation–reduction potential (ORP) by electrodes has been widespread for many years and the technology of such electrodes is thus well established. pH is measured by the potential produced by a glass electrode – an electrode with a special sensitive glass area and an acid electrolyte, used in conjunction with a standard calomel reference electrode. The output from the pH electrode is fed to an amplifier and then to a meter or digital display. A wide range of pH electrodes is available, including combined glass and reference units and special rugged units for field use. ORP is measured using a redox probe with a platinum electrode in conjunction with a calomel reference electrode.

More recent developments in electrode technology have resulted in the availability of a widening range of other electrodes, some of which are extremely useful in water quality control. Probably the most useful of these new electrodes is the oxygen electrode. These DO electrodes come in a number of configurations using lead/silver, carbon/silver or gold/silver cells, sheathed with a polythene film. Polythene is permeable to oxygen so that oxygen in the sample enters the cell and alters its electrical output in proportion to the oxygen concentration. An increasing number of specific ion electrodes for determinations such as NH_4^+, NO_3^-, Ca^{2+}, Na^+, Cl^-, Br^-, F^-, etc. are now available. These electrodes permit rapid measurements down to very low concentrations but they are relatively costly. Fouling of the sensitive surfaces by biological growths and calibration drift are often major operational problems with electrode sensors.

3.3 Automated analysis, remote monitoring and sensing

There has been a growing trend in the past decade to concentrate analytical facilities in central laboratories and this applies in the water industry as much as it does in the medical sciences where many of these techniques were originally developed. The growth of central laboratories capable of processing hundreds of thousands of samples a year has been encouraged by the availability of reliable automated analytical systems. These systems can carry out a suite of determinations on a series of samples without human intervention once the samples have been loaded into the system. Results are logged on computers and archived or reported as required. Many of the automated analytical systems utilize colorimetric determinations with automatic samplers feeding discrete samples into the reagent addition and colour development stages before entering the spectrophotometer flow-through cell. Individual metals can be determined to concentrations of $1\,\mu g$ or less using automated atomic absorption spectrophotometers in which atoms are excited in a flame or electric arc to produce a characteristic emission. In some analyses X-ray fluorescence spectroscopy may be used to improve sensitivity or to determine individual metals separately in a mixture. Chromatography is widely employed in the detection and identification of individual organic compounds. It utilizes the concept of differential absorption of organic compounds to separate individual components in mixtures. Gas–liquid chromatography, in which volatile substances are released by heating, is particularly useful for trace organics in water. For quantitative determinations of organic substances in the ng/l range, as required by drinking water standards, gas–liquid chromatography can be combined with mass spectrometry in which molecules are converted to ions and then separated by their mass:charge ratio. This enables identification and quantification of atoms and isotopes to provide a complete chemical analysis of a sample. High-technology analytical equipment such as that described above normally requires a controlled environment for it to operate reliably and it also needs skilled operators and maintenance technicians.

In pollution control work there is a need for on-site determinations of important quality parameters, often in remote situations with inhospitable environments. Equipment to fulfil such needs is required to operate for extended periods without attention other than routine cleaning and calibration. The attraction of probes or electrode sensors for such duties is obvious and much development work has been carried out by both instrument manufacturers and users. The production of reliable instruments has not been easy but they are now becoming available. The UK National Rivers Authority developed an automatic monitoring device which provides continuous records of dissolved oxygen, ammonia, pH, temperature and flow using probes which are automatically cleaned. Data are logged and telemetered to a control centre and a discrete sampler can be actuated automatically if preset levels are exceeded, thus providing samples for further investigation and legal action if appropriate. It is unlikely that it will ever be possible to provide remote monitoring of a water for all possible contaminants which might be present. In relation to raw water supplies it is, however, very desirable to have some form of automatic warning of the presence of potentially toxic substances. To protect potable water supplies a number of devices have been developed which depend upon the effect of toxic substances on the metabolism or behaviour of living organisms kept in a test chamber through which the water flows. Nitrifying bacteria are sensitive to many toxic substances and thus if exposed to them are likely to stop converting ammonia in the water into nitrate. An ammonia electrode placed at the outlet of a small column containing nitrifying bacteria will give rapid indication of a reduction in their activity. In a similar way, the respiration rate of fish, or in certain species changes in their electrical charge, can give early indication of stress due to undesirable substances in the water. It must be appreciated that such warnings simply indicate that the water has been contaminated and they cannot identify the particular contaminants. The warnings allow precautionary measures to be taken and must be followed up by the appropriate range of analyses to identify the substance causing the problem.

A major problem with conventional electrodes and probes is that they are inserted into the water or wastewater and thus as well as being prone to fouling and damage they may also alter the actual distribution of contaminants in the vicinity of the measurement. The concept of non-invasive sensors has thus begun to receive attention. It is already possible to obtain a turbidity monitor which utilizes light scattering from a falling stream of water as the measuring technique. The use of images from satellite sensors appears to have potential for some water quality monitoring purposes, particularly as the ground resolution of such sensors improves. Satellite images have been used to provide land-use data for catchment modelling purposes and to detect pollution discharges. Images can be subjected to a range of spectral analyses which can highlight particular characteristics and such images have already been used to detect algal blooms in large reservoirs. Flow measurements in rivers and open channels can now be made using non-intrusive ultrasonic beams and it may well be that in the not too distant future the reflection and scattering of laser beams could provide information about the quality of waters and wastewaters.

Further reading

Bartram, J. and Ballance, R. (1996). *Water Quality Monitoring*. London: E & F. N. Spon.

Briggs, R. and Grattan, K. T. V. (1990). The measurement and sensing of water quality: a review. *Trans. Inst. Meas. Control*, **12**, 65.

Ellis, J. C. and Lacey, R. F. (1980). Sampling: defining the task and planning the scheme. *Wat. Pollut. Control*, **79**, 452.

Finch, J., Reid, A. and Roberts, G. (1989). The application of remote sensing to estimate land cover for urban drainage catchment modelling. *J. Instn Wat. Envir. Managt*, **3**, 551.

Garry, J. A., Moore, C. J. and Hooper, B. D. (1981). Sewage treatment effluent sampling—have our means any meaning? *Wat. Pollut. Control*, **80**, 481.

Hey, A. E. (1980). Continuous monitoring for sewage-treatment processes. *Wat. Pollut. Control*, **79**, 477.

Hinge, D. C. (1980). Experiences in the continuous monitoring of river water quality. *J. Inst. Wat. Engnrs Scientists*, **34**, 546.

Hutton, L. G. (1983). *Field Testing of Water in Developing Countries*, Medmenham: Water Research Centre.

Ives, K. J. (1987). Sensing quality without touch. *Water Quality Intnl*, **4**(4), 30.

Morley, P. J. and Cope, J. (1980). Water quality monitoring. In *Developments in Water Treatment*, Vol. 2 (W. M. Lewis, ed.), p. 189. London: Applied Science Publishers Ltd.

Redfern, H. and Williams, R. G. (1996). Remote sensing: Latest developments and uses. *J. C. Instn Wat. Envir. Managt*, **10**, 423.

Schofield, T. (1980). Sampling of water and wastewater: practical aspects of sample collection. *Wat. Pollut. Control*, **79**, 477.

Shellens, M. and Edwards, Z. (1987). Optimization of sampling resources. *J. Instn Wat. Envir. Managt*, **1**, 297.

Standard methods of analysis

American Public Health Association (1995). *Standard Methods for the Examination of Water and Wastewater*, 19th edn. New York: APHA.

British Standards Institution (various dates). *Water Quality Determinations* (Separate booklets for individual determinations). London: BSI.

Department of the Environment (various dates). *Methods for the Examination of Waters and Associated Materials* (Separate booklets for individual determinations). London: HMSO.

Problems

1. In a total solids determination on a 100 ml sample of river water the following results were obtained:

Weight of empty dish after firing	56.3125 g
Weight of dish + sample after drying	56.3825 g
Weight of dish + sample after firing	56.3750 g

Determine the total and volatile solids concentrations in the water in mg/l. (700, 75)

2. Titration of a 50 ml sample of a groundwater with 0.02 N acid produced the following results:

Acid used to reach pH 8.2	6.1 ml
Additional acid used to reach pH 4.5	15.3 ml

Determine the total and caustic alkalinities and the forms in which the alkalinity occurs. (428 mg/l, 122 mg/l; CO_3^{2-} 244 mg/l, HCO_3^- 184 mg/l)

4

Aquatic microbiology and ecology

A feature of most natural waters is that they contain a wide variety of microorganisms forming a balanced ecological system. The types and numbers of the various groups of microorganisms present are related to water quality and other environmental factors. In the treatment of organic wastewaters, micro-organisms play an important role and most of the species found in water and wastewater are harmless to humans. However, a number of microorganisms are responsible for a variety of diseases and their presence in water creates a major health risk. It is therefore necessary to develop an understanding of the basic principles of microbiology and thus gain an appreciation of the role of microorganisms in water quality control.

4.1 Types of metabolism

Virtually all microorganisms require a moist environment for active growth but apart from this common feature many different types of metabolism are found. A basic classification can be made in relation to whether or not an organism requires an external source of organic matter.

Autotrophic organisms are capable of synthesizing their organic requirements from inorganic matter and can thus grow independently of external organics. Two methods may be employed to achieve this end, photosynthesis – many plants utilize inorganic carbon and ultraviolet radiation to produce organic matter and release oxygen

$$6CO_2 + 6H_2O \xrightarrow{\text{sunlight}} C_6H_{12}O_6 + 6O_2$$

and chemosynthesis – in which the chemical energy of inorganic compounds is utilized to provide the energy for synthesis of organics

$$2NH_3 + 3O_2 \rightarrow 2HNO_2 + 2H_2O + \text{energy}$$

Heterotrophic organisms require an external source of organic matter and are of three main types.

1. Saprophobes, which obtain soluble organic matter directly from the surrounding environment or by extracellular digestion of insoluble compounds. Food requirements can range from a simple organic carbon compound to a number of complex carbon and nitrogen compounds together with additional growth factors.
2. Phagotrophs, sometimes termed holozoic forms, can utilize solid organic particles.
3. Paratrophs obtain organic matter from the tissues of other living organisms and are thus parasitic.

Organisms differ in their requirements for oxygen – aerobes require the presence of free oxygen, whereas anaerobes exist in the absence of free oxygen. Facultative forms have a preference for one form of oxygen environment but can live in the other if necessary. In terms of temperature requirements there are three main types of organisms: psychrophilic, which live at a temperature close to 0°C; mesophilic, by far the most common, living within the range 15–40°C and thermophilic in the range 50–70°C. In practice there is a certain amount of overlap between these temperature ranges so that some organisms will be found growing actively at any temperature between 0 and 70°C.

4.2 Nomenclature

The nomenclature used in biology appears complex to the non-biologist but is very necessary because of the multiplicity of organisms which exists. A specific type of organism is denoted by its specific name and a collection of similar species is given a generic name, e.g. *Salmonella paratyphi*, a member of the *Salmonella* genus – the bacteria specifically responsible for paratyphoid; *Entamoeba histolytica*, the amoebic protozoan responsible for amoebic dysentery.

Drifting aquatic microorganisms are collectively termed phytoplankton if of plant origin and zooplankton when of animal origin. Free-swimming groups are called nekton, surface-swimming types are called neuston and bottom-living groups are referred to as benthos.

4.3 Types of microorganism

By definition, microorganisms are those organisms too small to be seen by the naked eye and there are large numbers of aquatic organisms in this category. With higher organisms it is convenient to identify them as plants or animals. Plants have rigid cell walls, are photosynthetic and do not move independently. Animals have flexible cell walls, require organic food and are capable of independent movement. The application of such differentiation to microorganisms is difficult because of the simple structures of their cells and it has become convention to term all microorganisms protists. The protists can themselves be divided into two types, prokaryotes and eukaryotes. Prokaryotes are small (<5 μm) simple cell structures with a rudimentary nucleus and one chromosome. Reproduction is normally by binary fission. Bacteria, actinomycetes and the blue-green algae are included in this group. Eukaryotes are larger (>20 μm) cells with a more complex structure and containing many chromosomes. Reproduction may be asexual or sexual and quite complex life cycles may be found. This class of microorganisms includes fungi, most algae and the protozoa.

There is a further group of microorganisms, the viruses, which do not readily fit into either of the above classes and which are thus considered separately.

Viruses

Viruses are the most basic form of organism, ranging in size from about 0.01 to 0.3 μm and they consist essentially of nucleic acid and protein. They are all parasitic and cannot grow outside another living organism. All are highly specific both as regards the host organism and the disease which they produce. Human viral diseases include smallpox, infectious hepatitis, yellow fever, poliomyelitis and a variety of gastrointestinal diseases. Because of the inability of viruses to grow outside a suitable host they are on the borderline between living matter and inanimate chemicals. Identification and enumeration of viruses requires special apparatus and techniques. Sewage effluents normally contain significant numbers of viruses, which are also present in most surface waters subject to pollution. Because of their small size their removal in conventional water-treatment processes cannot be certain although the normal disinfection processes will usually inactivate viruses.

Bacteria

Bacteria are single-cell organisms which utilize soluble food and may operate either as autotrophs or as heterotrophs. They range in size from 0.5 to 5 μm and have the basic features shown in Figure 4.1. Reproduction is by binary fission and the generation time may be as short as 20 minutes in favourable conditions with some species. Some bacteria can form resistant spores which can remain dormant for long periods in unsuitable environmental conditions but which can be reactivated on the return of suitable conditions. Most bacteria prefer more or less neutral pH although some species can exist in highly acid or highly alkaline environments. Bacteria play a vital role in the natural stabilization processes and are widely utilized for the treatment of organic wastewaters. There are some 1500 known species which are classified in relation to criteria such as size, shape and grouping of cells; colony characteristics; staining behaviour; growth require-ments; motility, specific chemical reactions; and so on. Aerobic, anaerobic and facultative forms are found. Most bacteria are harmless and even beneficial in the

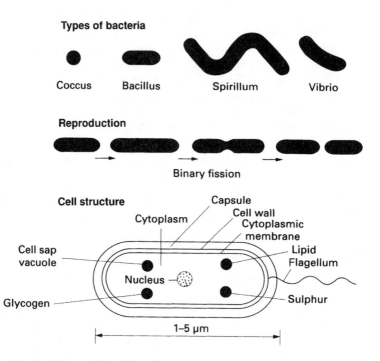

Figure 4.1   Basic characteristics of bacteria.

environment but some species are responsible for infections in humans and other animals. A wide range of gastrointestinal diseases is caused by bacterial contamination of water supplies or foods.

Fungi

Fungi are aerobic multicellular organisms which are more tolerant of acid conditions and a drier environment than bacteria. They utilize much the same food sources as the bacteria in chemosynthetic reactions but, because their protein content is somewhat lower than the bacteria, their nitrogen requirement is less. Fungi form rather less cellular matter than bacteria from the same amount of food. They are capable of degrading highly complex organic compounds and some are pathogenic in humans. Over 100 000 species of fungi exist and they usually have a complex structure formed of a branched mass of thread-like hyphae (Figure 4.2). They have four or five distinct life phases with reproduction by asexual spores or seeds. Fungi occur in polluted water and in biological treatment plants, particularly in conditions with high C:N ratios. They can be responsible for tastes and odours in water supplies.

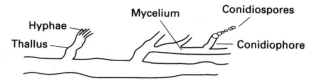

Figure 4.2   Main features of fungi.

Actinomycetes

The actinomycetes are similar to fungi in appearance with a filamentous structure but with a cell size close to that of bacteria. They occur widely in soil and water and nearly all are aerobic. Their significance in water is mainly due to the taste and odour problems which often result from their presence.

Algae

Algae are all photosynthetic plants, mostly multicellular although some types are unicellular. The majority of freshwater forms utilize the pigment chlorophyll and they act as the main producers of organic matter in an aquatic environment. Inorganic compounds such as carbon dioxide, ammonia, nitrate and phosphate provide the food source to synthesize new algal cells and to produce oxygen. In the absence of sunlight, algae operate on a chemosynthetic basis and consume oxygen so that in a water containing algae there will be a diurnal variation in DO levels; supersaturation may occur during the day, with significant oxygen depletion occurring at night. A large number of algae are found in freshwater and a variety of classification systems exists. Algae may be green, blue-green, brown or yellow depending upon the proportions of particular pigments. They occur as single cells which may be motile with the aid of flagella or non-motile, or exist as multicellular filamentous forms (Figure 4.3). Algae and bacteria growing in

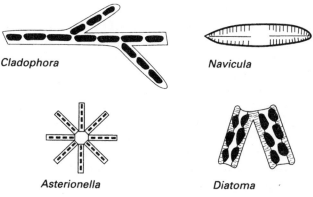

Figure 4.3   Some typical algae.

the same solution do not compete for food but have a symbiotic relationship (Figure 4.4) in which the algae utilize the end products of bacterial decomposition of organic matter and produce oxygen to maintain an aerobic system. In the absence of organic input, algal growth depends upon the mineral content of the water so that in a hard water algae obtain CO_2 from bicarbonates, reducing the hardness and usually increasing the pH, sometimes to high levels. Algae can be important in aquatic environments, particularly in standing or slowly moving bodies of water containing nutrients, phosphorus usually being the critical element, since many algae can fix atmospheric nitrogen. Many species of algae can cause taste and odour problems which become troublesome

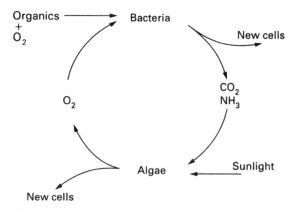

Figure 4.4 Symbiotic relationship between bacteria and algae.

during certain times of the year in waters used for public supply. Blue-green algae can grow prolifically in shallow lakes with quite small concentrations of phosphates. These algae release toxins which can be fatal to farm and domestic animals if they consume water which supports a significant algal population. The algal toxins can also cause skin irritation to humans in contact with the water and may produce gastrointestinal illness if the water is ingested. The increasing popularity of water-contact recreation has highlighted the need to monitor and control blue-green algae populations of the species *Microcystis, Apanizomenon, Anabena* and *Oscillatoria*.

Protozoa

The protozoa are unicellular organisms 10–100 μm in length which reproduce by binary fission. Most are aerobic heterotrophs and often utilize bacterial cells as their main food source. They cannot synthesize all the necessary growth factors and rely on the bacteria to provide these items. The protozoa are widespread in soil and water and may sometimes play an important role in biological waste-

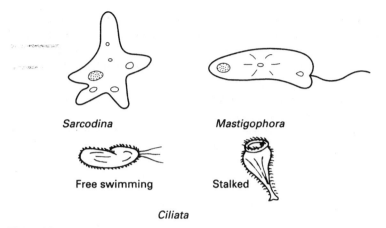

Figure 4.5   Some typical protozoa.

treatment processes. There are four main types of protozoa (Figure 4.5): *Sarcodina* – amoeboid flexible cell structure with movement by means of extruded pseudopod (false foot); *Mastigophora* – utilize flagella for motility; *Ciliata* – motility and food gathering by means of cilia (hair-like feelers), may be free swimming or stalked; *Sporozoa* – non-motile spore-forming parasites not found in water. Some pathogenic protozoans (*Giardia* and *Cryptosporidium*) found in water are capable of forming spores or cysts which are highly resistant to common disinfectants and can thus be a source of waterborne infection even in temperate developed countries.

Higher forms of life

As well as the microorganisms, more complex macroorganisms, many visible to the naked eye, are found in natural waters. These include rotifers, which are multicellular animals with a flexuous body and cilia on the head to catch food and provide motility, and crustaceans which are hard-shelled multicellular animals. Both of these groups provide important food supplies for fish and are normally only found in good-quality waters since they are sensitive to many pollutants and to low DO levels. Macro-invertebrates such as worms and insect larvae are useful as biological indicators in assessing water quality. They can also be found in some biological treatment processes where they may be able to metabolize complex organics not readily broken down by other organisms.

4.4 Microbiological examination

Because of their small size, observation of microorganisms with the naked eye is impossible and in the case of the simpler microorganisms their physical features do not provide positive identification. With bacteria it is necessary to utilize their biochemical or metabolic properties to aid identification of individual species. Living specimens are often difficult to observe because they usually have little colour and thus do not stand out against the liquid background. Staining of dead specimens is useful in some cases but the staining and mounting technique may itself alter some of the cell characteristics.

An optical microscope has a maximum magnification of about 1000 × with a limit of resolution of about 0.2 μm. Hence most viruses will be invisible with an optical microscope and only limited detail of the structure of a bacterial cell can be observed. For more detailed examination it is necessary to use an electron microscope which can provide magnifications of around 50 000 × with a limit of resolution of about 0.01 nm.

Study of living specimens, which is only possible under an optical microscope, is necessary to determine whether or not an organism is motile. A hanging drop slide must be used in this case and it is important to differentiate between Brownian movement, a random vibrational motion common to all colloids, and true motility which is often seen as a rapid shooting or wriggling movement.

In many situations it is necessary to assess the number of microorganisms present in a water sample. With the larger microorganisms like algae, estimation of numbers and actual identification of species can be achieved using a special microscope slide containing a depression of known volume. Numbers are obtained by counting the appropriate microorganisms in the chamber, aided by a grid etched on the slide.

An estimate of the number of living bacteria (viable cell count) in a water sample may be obtained with a plate count using nutrient agar medium. A 1 ml sample of the water, diluted if necessary, is mixed with liquefied agar at 40°C in a petri dish. The agar sets to a jelly thus fixing the bacterial cells in position. The plate is then incubated under appropriate conditions (72 h at 22°C for natural water bacteria, 24 h at 37°C for bacteria originating from animals or humans). At the end of incubation the individual bacteria will have produced colonies visible to the naked eye and the number of colonies is assumed to be a function of the viable cells in the original sample. In practice such plate counts do not give the total population of a sample, since no single medium and temperature combination will permit all bacteria to reproduce. However, viable counts at the two temperatures using a wide-spectrum medium do give an overall picture of the bacterial quality and pollution history of a sample.

To determine the presence of a particular genus or species of bacteria it is necessary to utilize its characteristic behaviour by supplying special selective media and/or incubation conditions which are suitable only for the bacteria under investigation. As is described in Chapter 5, many serious diseases are related to microbiological contamination of water, most of them due to pathogenic bacteria excreted by people suffering from or carrying the disease. Whilst it is possible to examine a water for the presence of a specific pathogen a more sensitive test employs the indicator organism *Escherichia coli* which is a normal inhabitant of the human intestine and is excreted in large numbers. Its presence in water thus indicates human excretal contamination and the sample is therefore potentially dangerous in that pathogenic faecal bacteria might also be present. Coliform bacteria in general have the ability to ferment lactose to produce acid and gas. Detection of coliforms can be achieved using a lactose medium in multiple tubes inoculated with serial dilutions of the sample. The appearance of acid and gas after 24 h at 37°C is taken as positive indication of the presence of coliform bacteria, results being expressed with the aid of statistical tables as most probable number (MPN)/100 ml. As a confirmatory test for *E. coli*, positive tubes are subcultured in fresh medium for 24 h at 44°C under which conditions only *E. coli* will grow to give acid and gas.

An alternative technique for bacteriological analysis, which has now largely replaced the multiple tube technique, uses special membrane filter papers with a pore size such that bacteria can be separated from suspension. The bacteria retained on the paper are then placed in contact with an absorbent pad containing the appropriate nutrient medium in a small plastic Petri dish and incubated. Identification of a particular species of bacteria is made on the basis of the type of nutrient provided and often on the appearance (colour, sheen) of the colonies formed. Counting the colonies provides the necessary quantitative information. The membrane filter technique is very convenient for laboratory and field testing although the cost of the materials is likely to be higher than for the multiple tube method of analysis unless the filters are washed for re-use. The numbers of microorganisms counted by the two techniques are not necessarily the same.

The identification and enumeration of specific pathogens in water is not easy and attempts are being made to automate such investigations. One possibility, which can be used to find *Cryptosporidium* oocysts involves adsorption of a fluorescent dye on the wall of living cells and detection of its presence by automated fluorescent cell cytology. Another technique under development for *Cryptosporidium* oocysts uses electro-rotation assay in which the sample is subjected to a rotating electrical field in which the oocysts exhibit specific spin characteristics differentiating between active and inactive specimens.

4.5 Ecological principles

In all communities of living organisms the various forms of life are interdependent to a greater or lesser extent. This interdependence is essentially nutritional, described as a trophic relationship, and is exemplified by the cycle of organic productivity and the carbon and nitrogen cycles. A biological community and the environment in which it is found form an ecosystem and the science of such systems is known as ecology.

The autotrophs in an ecosystem, i.e. green plants and some bacteria, are termed *producers* since they synthesize organic matter from inorganic constituents. Heterotrophic animals are known as *consumers* since they require ready-made organic food and may be subdivided into herbivores (plant eaters) and carnivores (meat eaters). Heterotrophic plants are termed *decomposers* since they break down the organic matter in dead plants and animals and in animal excreta. Some of the products of decomposition are utilized for their own growth and energy requirements, but others are released as simple inorganic compounds suitable for plant uptake. Figure 4.6 shows a simplified version of the carbon cycle. Solar radiation provides the only external energy source and permits the synthesis of carbohydrates and other organic products which are then transferred to the

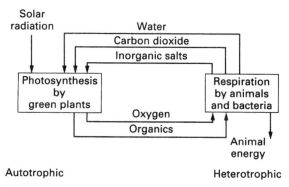

Figure 4.6 The carbon cycle.

heterotrophic phase of the cycle along with oxygen resulting from photosynthesis. In exchange, carbon dioxide, water and inorganic salts resulting from the activities of animals and bacteria are returned to the autotrophs. It should be noted that whilst carbon follows a cyclical path in such a system, energy flow is one-way only. It is important to remember that a continual energy input is thus necessary to allow the system to function. The loss of some part of the energy input to heat and entropy which inevitably occurs in biological systems can be considered as analogous to friction losses in a mechanical system. In fact the efficiency in terms of energy conversion of biological systems is very low and the further away an organism is from the original energy input the lower will be the portion of that energy available to the particular organism.

In an aquatic environment the interdependence of organisms takes the form of a complex food web within which are many food chains with successive links being composed of different species in a predator–prey relationship with adjacent links. Thus a typical food chain for a river would be

algae → rotifer → mayfly → minnow → pike

The successive links in the food chain contain fewer but larger individual organisms and the community can be pictured in the form of an Eltonian pyramid

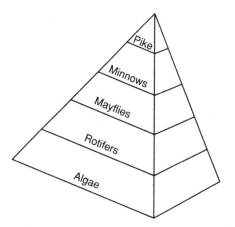

Figure 4.7   An Eltonian pyramid of numbers.

of numbers (Figure 4.7). Each level in the pyramid is known as a trophic level. Organisms occupying the same level compete for a common food, but those on a higher level are predatory on the lower level. Under natural conditions such an ecosystem can remain dynamically balanced over long periods, but changes in water quality or other environmental factors can completely upset the balance. Toxic materials tend to give a particular percentage kill of the population regardless of the population density, whereas the effect of such factors as shortage of food is more likely to be significant with dense populations. A clean surface water will normally contain many different forms of life, but none will be dominant and the community will be well balanced. Serious organic pollution of the water would produce conditions unsuitable for most of the higher forms of life so that the community would become one of one or two simple life forms which would be present in very large numbers because of the absence of predators.

Because of the low efficiency of ecological processes the numbers of organisms at the first trophic level required to support an organism at the top of the pyramid becomes very large. As a result, food chains in nature do not contain more than five or possibly six trophic levels.

Further reading

Abel, P. D. (1996). *Water Pollution Biology*, 2nd edn. Windsor: Taylor and Francis.
Bellinger, E. G. (1992). *A Key to Common Algae*, 4th edn. London: CIWEM.
Deere, D., Veal, D., Fricker, C. and Vesey, G. (1997). Automating analytical microbiology, *Wat. Envir. Manager*, **2**(1), 13.
Gaudy, A. F. and Gaudy, E. T. (1980). *Microbiology for Environmental Engineers and Scientists*. New York: McGraw-Hill.
Hawkes, H. A. (1963). *The Ecology of Wastewater Treatment*. Oxford: Pergamon.

Macan, T. T. and Worthington, E. B. (1972). *Life in Lakes and Rivers*. London: Fontana.

Mara, D. D. (1974). *Bacteriology for Sanitary Engineers*. Edinburgh: Churchill Livingstone.

McKinney, R. E. (1962). *Microbiology for Sanitary Engineers*. New York: McGraw-Hill.

Mitchell, R. (1974). *Introduction to Environmental Microbiology*. Englewood Cliffs: Prentice Hall.

National Rivers Authority (1990). *Toxic Blue Green Algae*. Bristol: NRA.

Odum, E. P. (1971). *Fundamentals of Ecology*, 3rd edn. Philadelphia: W. B. Saunders.

Sterritt, R. M. and Lester, J. N. (1988). *Microbiology for Environmental and Public Health Engineers*, London: E. & F. N. Spon.

5

Water quality and health

Because of the essential role played by water in supporting human life it also has, if contaminated, great potential for transmitting a wide variety of diseases and illnesses. In the developed world water-related diseases are rare, due essentially to the presence of efficient water supply and wastewater disposal systems. However, in the developing world perhaps as many as 1.3 thousand million people are without safe water supply and almost 2 thousand million do not have adequate sanitation. As a result, the toll of water-related disease in these areas is frightening in its extent. Millions of people die each year as the consequence of unsafe water or inadequate sanitation and although exact information is difficult to obtain, WHO data give an indication of the magnitude of the problem

- each year over five million people die from water-related diseases
- two million of the annual deaths are of children
- in developing countries 80 per cent of all illness is water-related
- at any one time half of the population in developing countries will be suffering from one or more of the main water-related diseases
- a quarter of children born in developing countries will have died before the age of five, the great majority from water-related disease.

At any one time there are likely to be 400 million people suffering from gastroenteritis, 200 million with schistosomiasis, 160 million with malaria and 30 million with onchocerciasis. All of these diseases can be water-related although other environmental factors may also be important.

In the developed world there is concern about the possible long-term health hazards which may arise from the presence of trace concentrations of impurities in drinking water, particular attention being paid to potentially carcinogenic compounds. There are several other chemical contaminants, which may be naturally occurring or man-made, having known effects on the health of consumers. It is therefore important that the relationships between water quality and health be fully appreciated by the engineers and scientists concerned with water quality control.

5.1 Characteristics of diseases

Before considering the water-related diseases it is useful to outline briefly the main features of communicable diseases.

All diseases require for their spread a source of infection, a transmission route, and the exposure of a susceptible living organism. Control of disease is thus based on curing sufferers, breaking the transmission route and protecting the susceptible population. Engineering measures in disease control are essentially concerned with breaking the transmission route and medical measures are concerned with the other two parts of the infection chain. It is interesting to note that a former Director General of WHO suggested that the number of water taps per 100 population is a better indication of health than the number of hospital beds. He should perhaps have stipulated that the taps delivered safe water and added the level of sanitation provision in his comparison, but the basic concept is very true.

Contagious human diseases are those in which the pathogen spends its life in the human body and can only live a short time in the unfavourable environment outside the body. This type of disease is thus transmitted by direct contact, droplet infection or similar means. With non-contagious diseases the pathogen spends part of its life cycle outside the human body so that direct contact is not of great significance. Non-contagious diseases may involve simple transmission routes with extracorporeal development of the infective organism taking place in soil or water. In many cases, however, more complex transmission routes occur with requirements for an intermediate host as part of the development of the parasite. It is thus important that control measures are developed in full knowledge of the transmission patterns of the particular disease. When a disease is always present in a population at a low level of incidence it is termed *endemic*. When a disease has widely varying levels of incidence the peak levels are called *epidemics* and worldwide outbreaks are termed *pandemics*.

5.2 Water-related disease

There are about two dozen infectious diseases, shown in Table 5.1, the incidence of which can be influenced by water. These diseases may be due to viruses, bacteria, protozoa or worms and although their control and detection is based in part on the nature of the causative agent it is often more helpful to consider the water-related aspects of the spread of infection. In the context of diseases associated with water there has in the past been some confusion about the terminology applied. Bradley (1977) developed a more specific classification system for water-related diseases which differentiates between the various forms of infections and their transmission routes.

Table 5.1 The main water-related diseases

Disease	Type of water relationship	Estimated annual deaths
Cholera Giardiasis Infectious hepatitis Leptospirosis Paratyphoid Tularaemia Typhoid	Waterborne	4 million
Amoebic dysentery Bacillary dysentery Gastroenteritis	Waterborne or water-washed	1 million
Ascariasis Conjunctivitis Diarrhoeal diseases Leprosy Scabies Skin sepsis and ulcers Tinea Trachoma	Water-washed	Relatively few deaths but large numbers of cases
Dracunculiasis Schistosomiasis	Water-based	200 thousand
Malaria Onchocerciasis Sleeping sickness Yellow fever	Water-related insect vector	1 million

Waterborne disease

The commonest form of water-related disease and certainly that which causes most harm on a global scale includes those diseases spread by the contamination of water by human faeces or urine. With this type of disease, infection occurs as shown in Figure 5.1 when the pathogenic organism gains access to water which is then consumed by a person who does not have immunity to the disease. The majority of diseases in this category, cholera, typhoid, bacillary dysentery, etc., follow a classical faecal–oral transmission route and outbreaks are characterized by simultaneous illness amongst a number of people using the same source of water. It should be appreciated that although these diseases can be waterborne they can also be spread by any other route which permits direct ingestion of faecal matter from a person suffering from that disease. Poor personal hygiene of workers in food handling and preparation would provide an obvious infection route. The situation is further complicated in that some people may be carriers of diseases like typhoid so that although they exhibit no outward signs of the disease their excreta contain the pathogens. Screening for such a carrier state is often mandatory for

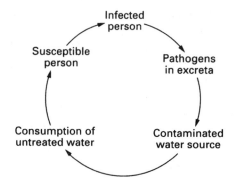

Figure 5.1 The classical waterborne disease infection cycle.

potential employees in the water-supply industry although medical opinion is divided as to whether the screening is really an effective precaution.

There are other waterborne diseases in which the infection pattern is not so simple. Weil's disease (leptospirosis) is transmitted in the urine of infected rats and the causative organism is able to penetrate the skin so that external contact with contaminated sewage or flood water can spread the disease.

Water-washed disease

As described above, infections which are transmitted by the ingestion of faecally contaminated water can also be spread by more direct contact between faeces and mouth. In the case of poor hygiene, due to inadequate water supply for washing, the spread of infection may be reduced by providing additional water, the quality of which becomes a secondary consideration. Water in this context is a cleansing agent and since it is not ingested the normal quality requirements need not be paramount. Clearly, waterborne faecal–oral diseases may also be classified as water-washed diseases and many of the diarrhoeal infections in tropical climates behave as water-washed rather than waterborne diseases.

A second group of diseases can also be classified as water-washed. These include a number of skin and eye infections which, whilst not normally fatal, have a serious debilitating effect on sufferers. The diseases of this type include bacterial ulcers and scabies, and trachoma. They tend to be associated with hot dry climates and their incidence can be significantly reduced if ample water is available for personal washing. This second group of water-washed infections is not also waterborne as those in the first group can be.

Water-based disease

A number of diseases depend upon the pathogenic organism spending part of its life cycle in water or in an intermediate host which lives in water. Thus human infection cannot occur by immediate ingestion of, or contact with, the organism excreted by a sufferer. Many of the diseases in this class are caused by worms which infest the sufferer and produce eggs which are discharged in faeces or urine. Infection often occurs by penetration of the skin rather than by consumption of the water.

Schistosomiasis (also called bilharzia) is probably the most important example of this class of disease. The transmission pattern of schistosomiasis is relatively complex in comparison with waterborne diseases and is illustrated in Figure 5.2. If a sufferer excretes into water, eggs from the worms hatch into larvae which can live for only 24 h unless they find a particular species of snail which acts as an intermediate host. The larvae then develop in a cyst in the snail's liver which, after about six weeks, bursts and releases minute free-swimming cercariae which

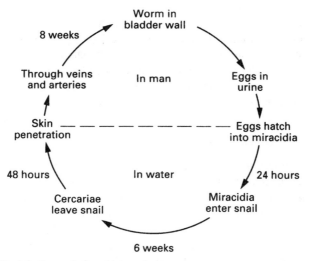

Figure 5.2 The infection cycle for schistosomiasis.

can live in water for about 48 h. The cercariae have the property of being able to puncture the skin of humans and other animals and they can then migrate through the body via skin, veins, lungs, arteries and liver in a period of around eight weeks. The parasite then develops in the veins of the wall of the bladder, or of the intestine, into a worm which may live several years and which will discharge enormous numbers of eggs.

It is unfortunate that schistosomiasis is often spread by irrigation schemes which, unless carefully designed and operated, tend to provide suitable habitats for the snail host as well as increasing the likelihood of contact with the water by agricultural workers. Control measures for schistosomiasis include, as well as the obvious prevention of excretal contamination of water, creation of conditions unfavourable to the presence of the snails, prevention of human contact with potentially contaminated water and the insertion of a 48 h delay period after removal of snails before access to the water is permitted. Unfortunately such measures are not easy to enforce and there may often be a conflict between the needs of schistosomiasis control and the desire to obtain agricultural and economic benefits from irrigation schemes in which some degree of water contact is inevitable.

Dracunculiasis (guinea worm) is another water-based disease which is widespread in the tropics. In this case the intermediate host is *Cyclops*, a small crustacean, and human infection occurs following the ingestion of water containing infected *Cyclops*. Eggs are discharged when an ulcer on the skin of a sufferer bursts and they can remain viable in water for one or two weeks. If eggs are ingested by *Cyclops* they develop into larval forms in a further two weeks. The larvae leave the ingested *Cyclops* during human digestive processes and migrate through the tissues to the lower limbs of the body and eggs are discharged about nine months later. The vector species of *Cyclops* is prevalent in stagnant water with some organic content. Control of guinea worm, which can bring marked improvements in the health of the population, is essentially based on protection of water sources, particularly springs and wells. The provision of sloping hardstandings and parapets round water sources will effectively prevent the access of eggs to the water.

Water-related insect vectors

There are a number of diseases that are spread by insects which breed or feed near water so that their incidence can be related to the proximity of suitable water sources. Infection with these diseases is in no way connected with human consumption of, or contact with, the water. Mosquitoes, which transmit malaria and a number of other diseases, prefer shallow stagnant water in pools, around the edges of lakes and in water storage jars. It is therefore important to ensure that water supply and drainage works do not provide suitable mosquito habitats, or, if this is unavoidable, mosquitoes should be prevented from gaining access by the provision of effective screens. *Simulium* flies, which transmit onchocerciasis (river blindness), breed in turbulent waters associated with rapids, waterfalls, etc., or created by engineering structures like weirs, energy dissipators, etc. Control is usually by use of insecticides injected upstream of the point of turbulence.

Water-related diseases in developed countries

The preceding text has been concerned with the major water-related diseases that are enormous hazards to public health in large areas of the world. Many of these diseases, particularly those of a gastrointestinal nature like cholera and typhoid, were widespread in Europe and North America last century but have virtually disappeared in developed countries because of improvements in public water supply and sanitation. However, the potential for waterborne disease is ever present and there have recently been a number of cases in the UK and in North America where intestinal infections have been caused by pathogenic protozoans present in public water supplies. These outbreaks have been caused by microorganisms of the *Cryptosporidium* and *Giardia* families, both of which can exist as small cyst forms which can be widespread in the environment since they occur in the faeces of many animals. The cysts are usually removed by sand filtration, but this cannot be guaranteed and they are resistant to chlorine (highly so in the case of *Cryptosporidium*) so that normal disinfection methods are unlikely to kill the organisms. Increasing concern at the possible health risks arising from these pathogens in water seem likely to give rise to some changes in the operation of rapid sand filters to prevent cysts appearing in the filtrate and the adoption of different disinfection techniques. The latter aspect is of particular importance when dealing with unfiltered surface waters. There is some concern that cryptosporidiosis may become more important because it appears to be especially hazardous to people who are immunosuppressed. A further problem with *Cryptosporidium* is that the infective dose appears to be quite low, perhaps twenty viable oocysts, compared with the infective dose for *Salmonella* species of around 2000 viable cells.

Outbreaks of Legionnaires' disease have lately come to some prominence and there is strong evidence to connect the incidence of this sometimes fatal type of pneumonia with the presence of *Legionella pneumophila* in domestic hot water supplies, shower heads, cooling waters and other aquatic systems which produce droplets or fine sprays. The disease is thus caused not by drinking microbiologically contaminated water but by breathing in a contaminated spray. The causative microorganism does in fact occur naturally in water sources, often those from underground, and since it is resistant to normal water treatment processes it can colonize water service systems in buildings, particularly those with warm surroundings. There is also some evidence that certain freshwater amoebae can enter the body in fine droplets of water and infect the respiratory tract. Such water-dispersed diseases are likely to be more of a problem in developed countries and where water-based recreations are popular. Water contact sports such as swimming, sail boarding and white-water canoeing, where immersion in the water and exposure to spray are likely, do appear to somewhat increase the chance of gastrointestinal and ear, nose and throat infections amongst participants.

5.3 Chemical-related illness

Because of the solvent properties of water many substances may be found in solution in natural waters and some of them are potentially hazardous to human life. Fortunately, the concentrations of most potentially harmful impurities in natural waters are normally very low but there are thousands of compounds used in agriculture, in the home and in industry which can find their way into surface and ground waters. In relation to chemically related health effects it is important to appreciate that these occur in two types: acute effects where the consequences of consumption of the contaminated water are more or less immediately apparent, and chronic effects where continued ingestion of the contaminated water produces a long-term hazard. The presence in drinking water of a concentration of a hazardous substance sufficient to cause an acute effect is only likely to occur as the result of an accident in the catchment area or in the treatment process. Either of these events would almost certainly result in appropriate preventative actions being taken by the authorities. In the event of such serious accidental contamination it would in any case be likely that the water would become unpalatable due to tastes and odours so that significant consumption would probably not occur. The main concern in water supply is thus the possible presence of low levels of contaminants which may produce observable health effects after long exposure, perhaps after many years. Some potentially harmful substances appear to have a threshold effect in that, provided the concentration remains below some critical level, no harm is caused. Other contaminants appear not to have a threshold level, so that any intake is potentially harmful and substances which cause cancer (carcinogens) often follow this pattern. The establishment of acceptable concentrations of contaminants in drinking water is particularly difficult in the case of chronic effects, since there is usually very little scientific evidence available to gauge the possible severity of the effects. The matter is further complicated by the fact that with many contaminants, drinking water is only one route of dietary intake and the ingestion of specific foods may be much more important. With threshold-type contaminants an acceptable daily intake can be computed using a body weight of 70 kg and a drinking water consumption of 2 litres/day and after allowing for intake in food and other drinks, which may be much greater than from drinking water. For other types of contaminants, particularly potential carcinogens such as polyaromatic hydrocarbons (PAH), trihalomethanes (THM), organochlorine and organophosphorus compounds and disinfection by-products (DBPs) it is advisable to ensure that concentrations in drinking water are kept as low as possible. This may mean that in some cases the allowable concentrations may be close to, or at, the limit of detectability. Such an approach embodies the precautionary principle rather than epidemiological evidence of actual hazard.

Lead

One of the earliest problems of chemical contamination of drinking water arose from the widespread use of lead piping and tanks in domestic systems. Soft acidic waters from upland catchments are often highly plumbosolvent so that significant amounts of lead can be dissolved in the water, particularly when standing in service pipes overnight. Lead is a cumulative poison so that continued exposure can eventually produce toxic effects and over the years most countries have made strenuous efforts to reduce the amount of lead in the environment. Lead-free fuels and paints are examples of this activity and attention has turned to reducing levels of lead in drinking water. The current EC drinking water standards set a maximum level of 50 µg/l but it seems likely that this will eventually be reduced to the 10 µg/l of the WHO guidelines, which will be difficult to achieve in areas with plumbosolvent water and lead service pipes. Chemical treatment of the water prior to distribution by the addition of lime or caustic soda reduces the dissolution of lead but removal of the lead piping would probably be essential to meet the 10 µg/l limit. Replacement of lead pipes will be costly to consumers since most of them are house connections and not the responsibility of the water supplier.

Nitrate

Nitrate nitrogen occurs naturally in many soils and is thus found in most groundwaters and in many surface waters. Nitrate concentrations in ground-waters have been rising in many developed countries because of the growth of intensive agricultural practices which utilize nitrogen fertilizers and alter natural drainage patterns. Sewage effluents also contain significant amounts of nitrogen which is often in the nitrate form. Although the presence of nitrates in drinking water has not been shown to be harmful to children or adults, it can be dangerous for bottle-fed babies up to the age of about six months. Below this age babies do not have the normal bacterial flora in their intestines and they are unable to deal with the nitrites produced by reduction of nitrate in the stomach. If a baby is bottle-fed with milk made up with water containing more than 10–20 mg/l of nitrate nitrogen there is a possibility of the nitrite absorbed in the blood preventing oxygen transport and causing methaemoglobinaemia, 'blue baby disease'. Although some groundwaters have always contained high levels of nitrate and there is a trend of increasing nitrate levels in many aquifers, very few cases of methaemoglobinaemia have occurred this century and most of these have been linked to high nitrate levels in private water supplies.

Fluoride

Fluoride occurs naturally in some waters and its presence in drinking water has been shown to be inhibitory to tooth decay, particularly when young children are exposed. As a result of this effect some health authorities recommend fluoridation of drinking water to provide a concentration of 1 mg/l. This recommendation is vociferously opposed in some quarters as an example of mass medication. At levels of fluoride above 1.5 mg/l there is a possibility of yellow staining of teeth and at much higher levels there is the danger of bone damage through fluorosis. Fluoride is thus an example of a constituent of water which can be beneficial at low concentrations but which becomes harmful at higher concentrations. Such effects are not uncommon with many chemicals found in the diet.

Aluminium

Aluminium occurs naturally in some raw water sources and it is commonly used as a coagulant in water treatment processes. In normal circumstances the aluminium is converted to an insoluble form and is thus removed from the water. There are, however, occasions where some soluble aluminium may be present in the water, although the amount ingested by this route is likely to be a small fraction of the total dietary intake. The use of aluminium cooking utensils, particularly for stewing acid fruit, can produce aluminium intakes much greater than those that would normally arise from aluminium in drinking water. Very high levels of aluminium can have a number of health consequences and an accidental discharge of concentrated aluminium sulphate to drinking water in Camelford south-west England in 1988 raised concerns about operational procedures. It has been suggested that the small doses of soluble aluminium which can be found in drinking water are in a chemical form which makes them subject to metabolism in the body in such a way as to increase the incidence of Alzheimer's disease, a form of senile dementia. Definitive proof of such an effect has not yet appeared, but the possibility of a connection has caused some water treatment plants to substitute iron salts for aluminium as a coagulant. It should, however, be noted that the 1993 WHO guidelines do not class aluminium as a health-related parameter. A specialized health hazard arising from the presence of aluminium in water is that patients on kidney dialysis machines can be fatally affected if the supply to the machine contains soluble aluminium.

Arsenic

In some areas, notably in parts of Argentina, Chile, China, India, Mexico and Taiwan, groundwaters may contain arsenic concentrations of up to several milligrammes per litre. The provisional WHO guideline level for arsenic is now 10 μg/l although the current EC MAC level is 50 μg/l. Regular consumption of water with higher concentrations of arsenic can lead to skin pigmentation and a variety of gastrointestinal, haematological and renal disorders.

Iodine

Certain elements are essential for life and in some situations a deficiency of an element in drinking water can cause health problems in a community. The best known example of this problem is that of iodine deficiency leading to endemic goitre and cretinism. As many as a billion people worldwide are at risk from iodine-deficiency disorders and most are in developing countries where poor and limited diets are likely to be contributory factors. Iodine-deficient waters usually arise in remote inland communities since coastal rainfall normally contains iodine picked up from its passage over the sea. In developed countries any risk of iodine deficiency is usually prevented by using iodized table salt but this is not usually available in those parts of developing countries which may be at risk.

Hardness

There is strong statistical evidence of a correlation between increased hardness in drinking water, up to about 175 mg/l, and reduced incidence of some forms of heart disease. This evidence would therefore suggest that softening of hard waters could have detrimental health effects. Some forms of softening increase the sodium content of the water and this can be undesirable for patients suffering from some heart and kidney complaints.

Reference

Bradley, D. J. (1977). Health aspects of water supplies in tropical countries. In *Water, Wastes and Health in Hot Climates* (R. G. Feachem, M. McGarry and D. D. Mara, eds). Chichester: Wiley.

Further reading

Ainsworth, R. G. (1990). Water treatment for health hazards. *J. Instn Wat. Envir. Managt*, **4**, 489.

Bragg, S., Soliars, C. J. and Perry, R. (1990). Water quality for renal dialysis. *J. Instn Wat. Envir. Managt*, **4**, 203.

British Geological Survey (1996). *Groundwater, Geochemistry and Health*. Wallingford: BGS.

Britton, A. and Richards, W. W. (1981). Factors influencing plumbosolvency in Scotland. *J. Instn Wat. Engnrs Scientists*, **35**, 349.

Cairncross, S. and Tayer, A. (1988). Guinea worm and water supply in Kordofan, Sudan. *J. Instn Wat. Envir. Managt*, **2**, 268.

Colbourne, J. S. and Dennis, P. J. (1989). The ecology and survival of *Legionella pneumophila*. *J. Instn Wat. Envir. Managt*, **3**, 345.

Craun, G. F. (1988). Surface water supplies and health. *J. Am. Wat. Wks Assn*, **80**(2), 40.

Craun, G. F., Swerdlow, D., Tauxe, R., *et al.* (1991). Prevention of waterborne cholera in the United States. *J. Amer. Wat. Wks Assn*, **83**(11), 40.

Dadswell, J. V. (1990). Microbiological aspects of water quality and health. *J. Instn Wat. Envir. Managt*, **4**, 515.

Feachem, R. G., Bradley, D. J., Garelick, H. and Mara, D. D. (1983). *Sanitation and Disease—Health Aspects of Excreta and Wastewater Management*. Chichester: Wiley.

Galbraith, N. S., Barrett, N. J. and Stanwell-Smith, R. (1987). Water and disease after Croydon: A review of water-borne and water-associated disease in the UK 1937–86. *J. Instn Wat. Envir. Managt*, **1**, 7.

Herwaldt, B. L., Craun, G. F., Stokes, S. L. and Juranek, D. D. (1992). Outbreaks of waterborne disease in the United States 1989–90. *J. Amer. Wat. Wks Assn*, **84**(4), 129.

Jones. F., Kay, D., Stanwell-Smith, R. and Wyer, M. (1990). An appraisal of the potential health impacts of sewage disposal to UK coastal waters. *J. Instn Wat. Envir. Managt*, **4**, 295.

Lacey, W. F. (1981). *Changes in Water Hardness and Cardiovascular Death Rates*, TR 171. Medmenham: Water Research Centre.

Lacey, R. F. and Pike, E. B. (1989). Water recreation and risk. *J. Inst Wat. Envir. Managt*, **3**, 13.

Le Chevallier, M. W. and Norton, W. D. (1995). *Giardia* and *Cryptosporidium* in raw and finished waters. *J. Amer. Wat. Wks Assn*, **87**(9), 54.

Neal, R. A. (1990). Assessing toxicity of drinking water contaminants: An overview. *J. Am. Wat. Wks Assn*, **82**(10), 44.

Overseas Development Administration (1996). *Water for Life*. London: ODA.

Packham, R. F. (1990). Chemical aspects of water quality and health. *J. Instn Wat. Envir. Managt*, **4**, 484.

Packham, R. F. (1990). Cryptosporidium and water supplies—the Badenoch report. *J. Instn Wat. Envir. Managt*, **4**, 578.

Pontius, F. W., Brown, G. and Chen, C.-J. (1994). Health implications of arsenic in drinking water. *J. Amer. Wat. Wks Assn*, **86**(9), 52.

Smith, H. V. (1992). *Cryptosporidium* and water. *J. Instn Wat. Envir. Managt*, **6**, 443.

World Health Organization (1992). *Our Planet, Our Earth*. Geneva: WHO.

6

Biological oxidation of organic matter

Many of the problems associated with water quality control are due to the presence of organic matter from natural sources or from wastewater discharges. This organic matter is normally stabilized biologically and the microorganisms involved utilize either aerobic or anaerobic oxidation systems.

In the presence of oxygen, aerobic oxidation takes place, part of the organic matter being synthesized to form new microorganisms and the remainder being converted to relatively stable end products as shown in Figure 6.1. In the absence of oxygen, anaerobic oxidation will produce new cells and unstable end products such as organic acids, alcohols, ketones, methane.

The anaerobic methane-producing system which is used in wastewater treatment takes place in two stages. In the first stage acid-forming micro-organisms convert the organic matter into new cells and organic acids and alcohols. A second group of microorganisms, the methane bacteria, then continue the oxidation again utilizing part of the organic matter to synthesize new cells and converting the remainder to methane, carbon dioxide and hydrogen sulphide. The anaerobic reaction is much slower than the aerobic process and is inefficient as regards energy conversion; aerobic oxidation of glucose, for example, yields about thirty times the energy released by its oxidation anaerobically.

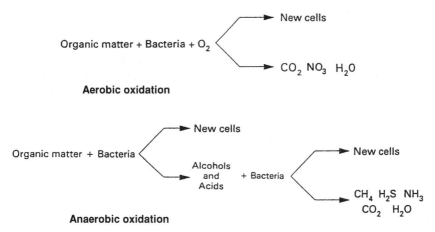

Organic matter + Bacteria + O_2 → New cells / CO_2 NO_3 H_2O

Aerobic oxidation

Organic matter + Bacteria → New cells / Alcohols and Acids + Bacteria → New cells / CH_4 H_2S NH_3 CO_2 H_2O

Anaerobic oxidation

Figure 6.1 Modes of biological oxidation.

6.1 Nature of organic matter

There are three main types of organic matter in water quality control.

1. Carbohydrates, containing carbon, hydrogen and oxygen (CHO). Typical examples are sugars, e.g. glucose, starch, cellulose.
2. Nitrogenous compounds, containing carbon, hydrogen, oxygen, nitrogen and occasionally sulphur (CHONS). The main compounds in this group are proteins which are very complex molecules, amino acids (the building blocks of proteins) and urea. The nitrogen in these compounds is liberated as ammonia on oxidation.
3. Lipids or fats, containing carbon, hydrogen and a little oxygen (CHO). They are only slightly soluble in water but soluble in organic solvents.

Because of the numerous organic compounds found in wastewaters it is rarely profitable or even feasible to isolate them. It is normally sufficient to determine the total amount of organic matter present (CHONS).

6.2 Biochemical reactions

Microorganisms use organic substances as a food source by means of a series of complex reactions. These reactions may be catabolic, in which the food is broken down to release energy, or anabolic, in which energy is utilized to synthesize new microbial cells. A vital link in the biochemical transfer of energy involves the compound adenosine triphosphate (ATP), the stored energy in which is released by breaking a phosphate bond and forming adenosine diphosphate (ADP). Energy from organic matter can then be used to convert ADP back to ATP to carry out further energy transfer reactions.

Biochemical reactions are controlled by enzymes which are organic catalysts, produced by living organisms, capable of speeding up the reactions without being consumed in the process. Enzymes are proteins of high molecular weights which are able to catalyse specific biochemical reactions. Their performance is influenced by such factors as temperature, pH, substrate concentration and the presence of inhibitors. There are many different types of enzyme, characterized by names ending in -*ase* (oxidase, dehydrogenase, etc.), and they are classified by reference to the reaction which they control. Important enzyme-catalysed reactions in water quality control are

- oxidation – addition of oxygen or removal of hydrogen
- reduction – addition of hydrogen or removal of oxygen
- hydrolysis – addition of water
- dehydrolysis – removal of water
- deamination – removal of NH_2 group.

Reactions which occur mainly outside the microbial cell are catalysed by hydrolytic extracellular enzymes located on the cell surface. Oxidation reactions occur when the substrate has been broken down by hydrolysis into units which are able to pass through the cell wall to come under the influence of intracellular enzymes.

6.3 Nature of biological growth

For successful biological growth certain requirements must be satisfied.

1. *Sources of carbon and nitrogen.* An empirical formula for a bacterial cell is $C_{60}H_{87}O_{23}N_{12}P$ and this has implications on the necessary composition of a wastewater if biological treatment is to be used. Thus carbon, nitrogen and, to a lesser extent, phosphorus are essential elements for growth. In practical terms 1 kg of organic and ammonia nitrogen is required for between 15 and 30 kg of BOD, and 1 kg of phosphorus is required for between 80 and 150 kg of BOD.
2. *Energy sources.* Microorganisms require energy for their metabolic activities and this energy must be made available by releasing the energy of formation bound up in chemical compounds when they were originally formed from their basic constituents.
3. *Inorganic ions.* Many inorganic ions, mainly metals such as calcium, magnesium, potassium, iron, manganese, cobalt, etc., are essential to growth, although they are only required in minute amounts. Such ions are normally present in the water supply and thus also in sewage.
4. *Growth factors.* Like other living organisms, microorganisms need growth factors such as vitamins, but here again these are normally present at the required level in wastewater.

The absence of sufficient amounts of any of the above items can significantly influence the biological treatability of a wastewater and in some cases it may be necessary to add the missing component to the wastewater. Vegetable processing wastes, for example, are often low in nitrogen so that ammonium salts may have to be added to ensure effective biological treatment.

In a system where an inoculum of bacteria is added to a suitable substrate containing all the necessary nutrients and growth factors, the classical biological growth curve as shown in Figure 6.2 will be reproduced. If the bacteria have not previously been grown in the particular substrate, there may be a lag period in which the microorganisms develop adaptive enzymes to enable them to utilize the substrate.

The initial active growth phase takes place with an excess of food so that growth is limited only by the rate at which the microorganisms can reproduce. In

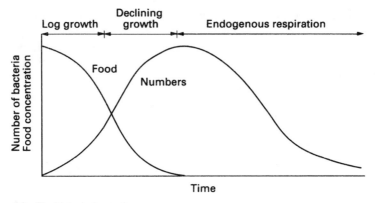

Figure 6.2 The biological growth curve.

such circumstances the growth rate is logarithmic and the rate of increase of microorganisms (or biomass) is proportional to the mass of microorganisms present in the system, i.e.

$$\frac{dX}{dt} = -\mu X \qquad (6.1)$$

where X = biomass concentration, t = time and μ = specific growth rate.
    Hence

$$X_t = X_0 \, e^{-\mu t} \qquad (6.2)$$

or

$$\log_e X_t = \log_e X_0 + \mu t \qquad (6.3)$$

where X_t = biomass concentration at time t and X_0 = biomass concentration at time $t = 0$.

    Thus by plotting experimental determinations of biomass against time on a semi-log paper the slope of the linear portion of the line gives the value of the specific growth rate μ.

    In the declining growth phase the rate of growth becomes increasingly limited by a reduction in the food concentration and possibly also by the accumulation of toxic end products in the system. When all the substrate has been exhausted, growth will cease and the numbers of microorganisms begin to fall in an autodigestion or endogenous respiration phase. Dead cells lyse and thus release some nutrients which provide a source of food for those organisms still living. Endogenous respiration occurs at an exponentially declining rate which is analogous to the decay of radioisotopes.

In this situation

$$\frac{dX}{dt} = - bX \qquad (6.4)$$

where b = endogenous respiration constant, or

$$X_t = X_0 \, e^{-bt} \qquad (6.5)$$

where X_0 = biomass concentration at start of endogenous respiration phase, $t =$ 0 and X_t = biomass concentration at time t after start of endogenous respiration phase.

Not all of a microbial cell is capable of being utilized by other micro-organisms, so that in the presence of endogenous respiration it is important to appreciate that the relationship between the mass of living cells and the mass of biological material will not remain constant.

When considering wastewater treatment there is obviously a desire to reduce the substrate concentration to as low a value as possible. However, it is clear from Figure 6.2 that the time of complete removal of substrate corresponds to the largest biomass production which can in practical terms be equated to sludge production. Some reduction in sludge mass can be achieved by permitting endogenous respiration to take place, but this carries the penalty of a longer reaction time. This implies a larger treatment plant and thus increased operating costs. It is important to appreciate that the curve shown in Figure 6.2 is based on a batch reaction, whereas most treatment processes operate in a continuous mode with the reaction being fixed at a point on the time axis. It is useful to have some measure of the sludge producing potential of a substrate and this is given by the yield coefficient Y,

$$Y = \frac{(X_t - X_0)}{(S_0 - S_t)} \qquad (6.6)$$

where S_0 and S_t represent substrate concentrations at times 0 and t respectively with X_0 and X_t being the biomass concentrations at these times. The actual value of the yield coefficient is influenced by the nature of the substrate and the type of microorganisms in the process as well as other environmental factors. For single substrates Y may vary from 0.29 to 0.70, but in relation to wastewaters, which are usually complex substrates, it is convenient to express Y as mass of volatile suspended solids (VSS) produced per mass of oxygen demand removed. In aerobic wastewater treatment systems a Y value of 0.55 in terms of COD is often used and for anaerobic systems the appropriate value of Y is usually about 0.1.

In the earlier discussion of log growth it was postulated that the specific growth rate was unaffected by substrate concentration. In the limit this is no longer true and the declining growth phase comes into play. In any system it would obviously be useful to be able to predict the substrate concentration which becomes growth limiting. Experimental studies of the relationship between specific growth rate and substrate concentration show that in a pure-culture single-substrate system the specific growth rate approaches its maximum value asymptotically as the substrate concentration increases. This concept gives rise to the Michaelis–Menten equation

$$\mu = \frac{\mu_m S}{(K_s + S)} \tag{6.7}$$

where μ_m = maximum specific growth rate, S = substrate concentration and K_s = substrate concentration which gives $\mu = 0.5 \, \mu_m$.

This equation is useful for modelling simple reactions, but for complex substrates with mixed cultures of microorganisms, like most wastewaters, its validity is somewhat questionable.

6.4 Oxygen demand in aerobic oxidation

It is of great importance in water quality control that the amount of organic matter present in the system be known and that the quantity of oxygen required for its stabilization be determined. In the case of a simple compound like glucose it is possible to write down the equation for its complete oxidation

$$C_6H_{12}O_6 + 6O_2 \rightarrow 6CO_2 + 6H_2O$$

i.e. each molecule of glucose requires six molecules of oxygen for complete conversion to carbon dioxide and water. In the case of the more complex compounds found in most samples, e.g. proteins, etc., the reactions become more difficult to understand. In addition to the oxygen required to stabilize carbonaceous matter, there is also a considerable oxygen demand during the nitrification of nitrogenous compounds

$$2NH_3 + 3O_2 + \text{nitrifying bacteria} \rightarrow 2NO_2^- + 2H^+ + 2H_2O$$

$$2NO_2^- + O_2 + 2H^+ + \text{nitrifying bacteria} \rightarrow 2NO_3^- + 2H^+$$

The amount of oxygen required to stabilize a waste completely could be calculated on the basis of a complete chemical analysis of the sample, but such a determination would be difficult and time consuming. Several methods of calculating the theoretical oxygen demand knowing various characteristics of the sample have been proposed, e.g.

$$\begin{aligned} \text{Ultimate oxygen demand} &= 2.67 \times \text{Organic carbon (mg/l)} + \\ \text{(UOD) (mg/l)} \quad & 4.57(\text{Org. N} + \text{Amm. N}) \text{ (mg/l)} + \\ & 1.14 \times \text{Nitrite N (mg/l)} \end{aligned} \tag{6.8}$$

The chemical oxygen demand determinations using potassium permanganate or potassium dichromate measure a proportion of the UOD (a large proportion in the case of the dichromate method). Unfortunately these methods have two basic disadvantages in that they give no indication of whether or not the substance is degradable biologically and nor do they indicate the rate at which biological oxidation would proceed and hence the rate at which oxygen would be required in a biological system. Because of these disadvantages a great deal of work on waste strength measurements is still done using the biochemical oxygen demand (BOD) test developed by the Royal Commission on sewage disposal at the turn of the century.

The BOD test measures the oxygen consumed by bacteria whilst oxidizing organic matter under aerobic conditions. The oxidation proceeds relatively slowly and is not usually complete in the standard five day period of incubation. Simple organic compounds like glucose are almost completely oxidized in five days, but domestic sewage is only about 65 per cent oxidized and complex organic compounds might be only 30–40 per cent oxidized in this period. Exertion of BOD is normally assumed to follow a first-order reaction initially, although there is evidence that biological oxidation in practice is not necessarily so straightforward. In a first-order reaction the rate of oxidation is proportional to the concentration of oxidizable organic matter remaining and once a suitable population of microorganisms has been formed the rate of reaction is controlled only by the amount of food available, i.e.

$$\frac{dL}{dt} = -KL \tag{6.9}$$

where L = concentration of organic matter remaining, or ultimate BOD, t = time and K = constant specific to the particular organic substance or substances present.

Integrating,

$$\frac{L_t}{L} = e^{-Kt} \tag{6.10}$$

where L_t is the BOD remaining at time t, as indicated in Figure 6.3.

It is conventional to use $\log_{10}$ rather than $\log_e$ and this can be achieved by changing the constant

$$\frac{L_t}{L} = 10^{-kt} \tag{6.11}$$

where $k = 0.4343\,K$ and k is termed the rate constant.

The normal concern is with oxygen taken up, i.e. BOD, rather than with oxygen demand remaining, thus

$$\text{BOD}_t = (L - L_t) = L(1 - 10^{-kt}) \tag{6.12}$$

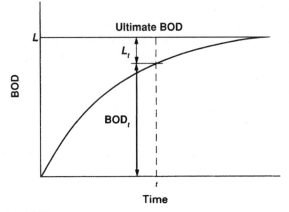

Figure 6.3   Basis of BOD.

The value of k governs the rate of oxidation, as shown in Figure 6.4, and may be used to characterize the biological degradability of a substance. For domestic sewage, k is about 0.17/day at a temperature of 20°C. For another temperature T new values can be found from

$$k_T = k_{20}(1.047)^{(T-20)} \tag{6.13}$$

$$L_T = L_{20}[1 + 0.02(T - 20)] \tag{6.14}$$

The determination of a single five day BOD does not of course permit calculation of L and K values for the sample. To obtain this information it is

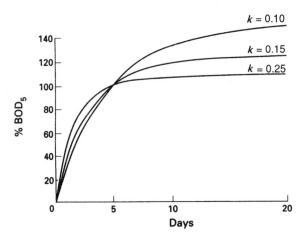

Figure 6.4   Effects of k values on oxygen demand.

necessary to carry out determinations of BOD over periods of, say, one to six days and then subject the data to some form of curve fitting process. A simple fitting technique due to Thomas (1950) relies on the fact that the expansions of $(1 - e^{-Kt})$ and $Kt(1 + Kt/6)^{-3}$ are very similar. The expression $BOD_t = L(1 - e^{-Kt})$ can thus be approximated by $BOD_t = LKt(1 + Kt/6)^{-3}$, i.e.

$$\left(\frac{t}{BOD_t}\right)^{1/3} = (KL)^{-1/3} + \left(\frac{K^{2/3}}{6L^{1/3}}\right) t \qquad (6.15)$$

By plotting $(t/BOD_t)^{1/3}$ against t a straight line should be produced with the intercept $a = (KL)^{-1/3}$ and the slope $b = K^{2/3}/6L^{1/3}$. Hence

$$K = \frac{6b}{a} \qquad (6.16)$$

and

$$L = \frac{1}{Ka^3} \qquad (6.17)$$

In practice the shape of the BOD curve is modified by the effect of oxygen required for nitrification (Figure 6.5), but due to the slow rate of growth of nitrifying bacteria this effect is not normally important until eight to ten days have elapsed with raw waste samples. However, in the case of treated effluents the effect of nitrification may become apparent after a day or two due to the presence of large numbers of nitrifying bacteria in the effluent. Nitrification can be inhibited in BOD samples by the addition of allyl thiourea (ATU), so that only carbonaceous oxygen demand is measured. It should be remembered, however, that nitrification does exert a significant oxygen demand. It is sometimes argued that the presence of nitrates in an effluent provides an oxygen reservoir which

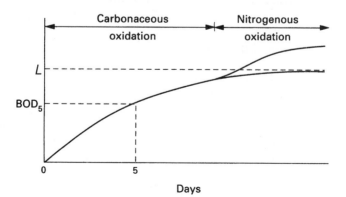

Figure 6.5   Typical BOD curve.

may be utilized if further pollution occurs in the river. Unfortunately, however, oxygen from nitrates is only released when the DO falls below 1.0 mg/l, by which time much of the life in the water has been killed. Because of the biochemical nature of the BOD determination its reproducibility is rarely better than ±10 per cent and experimental results will only produce an approximation to a linear relationship when transformed in the manner described.

The BOD test was originally intended to simulate conditions which occurred following the discharge of an effluent to a river, and whilst there may be some resemblance between the conditions of the test and conditions in a surface water there is little relation between the test conditions and those which prevail in a biological treatment plant. The BOD test uses a small culture of microorganisms to stabilize organic matter in quiescent conditions and constant temperature with a limited DO supply. In a biological treatment plant high concentrations of microorganisms are continuously agitated to keep them in contact with the concentrated substrate and an excess of DO is supplied. A further disadvantage of the BOD determination is that the results give no indication of the rate of oxygen uptake unless BODs are determined at daily intervals over a period instead of the standard five day period.

Because the solubility of oxygen in water at 20°C is only 9.1 mg/l many samples would utilize all the DO long before the five days had elapsed. To determine BOD values on most samples it is therefore necessary to estimate the likely result and then dilute the sample by an amount sufficient to leave 1–2 mg/l DO at the end of the test period. The diluted sample and the dilution water are aerated and their DO concentrations checked before BOD bottles are filled and incubated for five days. The remaining DO concentrations in the diluted sample and dilution water bottles are determined and then BOD calculated as shown in the example below.

Worked example for BOD determination

An industrial wastewater sample was diluted to 1 in 5 and the DO results obtained were

initial DO of diluted sample	9.10 mg/l
final DO of diluted sample	4.30 mg/l
initial DO of dilution water	9.10 mg/l
final DO of dilution water	8.70 mg/l

The oxygen consumption of the diluted sample (9.10 – 4.30 mg/l) is due to 1/5 of the bottle containing wastewater with the remaining 4/5 being dilution water. The dilution water itself will have a small BOD which in a full bottle is (9.10 – 8.70). The BOD contribution of 4/5 of a bottle of dilution water must therefore be deducted from the oxygen consumption in the diluted sample and the result then multiplied by the dilution factor to give the undiluted BOD of the wastewater.

Thus for the wastewater

$$\begin{aligned} \text{BOD} &= [(9.10 - 4.30) - (9.10 - 8.70)\ 4/5] \times 5 \\ &= [4.80 - (0.40 \times 4/5)] \times 5 \\ &= 4.48 \times 5 = 22.40\,\text{mg/l} \end{aligned}$$

In attempts to obtain a more realistic measurement of oxygen uptake in the conditions which prevail in biological wastewater treatment plants, i.e. high food concentrations and large populations of microorganisms, respirometers have had some success. A simple respirometer comprises a sealed flask in which a suitable sample of biomass and food source are placed and kept under agitation by a magnetic stirrer. The carbon dioxide produced by respiration of the micro-organisms is absorbed with a caustic solution so that there is a resulting drop in pressure in the flask which can be measured by a manometer. More sophisticated respirometers utilize DO electrodes to give a continuous measurement of the oxygen concentration in the sample and hence enable calculation of the oxygen uptake rate. Another type uses the pressure drop to energize an electrolytic oxygen generator which introduces sufficient oxygen into the flask to return the pressure to a standard value. The length of time for which the oxygen generator operates can be converted into a measure of oxygen transfer. Whilst respirometers are of considerable value in research and possibly in process control, the BOD is still the standard measure of oxygen demand for most purposes.

It is vital to appreciate that any biochemical test may give erroneous results in the presence of inhibitory or toxic materials. With unfamiliar samples it is therefore prudent to determine COD values in parallel with the BOD determinations. Substances with COD/BOD ratios of between 1 and 3 are likely to be reasonably biodegradable, but if the ratio is greater than 5, suspicions should be aroused. In such cases it may be that the organic matter is not readily biodegradable, at least without a period of acclimation, or the presence of other ingredients may be preventing or inhibiting biological oxidation. In either situation it is important to undertake further investigations to determine the cause of the discrepancies between the COD and BOD values. It should be noted that as biochemical oxidation of a biodegradable sample proceeds, the ratio of COD/BOD will increase because refractory or non-biodegradable substances contribute a larger proportion of the COD of the sample. Thus a wastewater with an initial COD/BOD ratio of 2.5 might have a ratio of 6 or 7 after intensive aerobic treatment. As discussed in Chapter 7, this situation can give problems if COD limits are included in effluent discharge regulations.

6.5 Anaerobic oxidation

With certain strong organic wastes, e.g. sludges, slaughterhouse discharges, etc., the oxygen requirement for aerobic stabilization is high and it becomes physically difficult to maintain aerobic conditions in the reaction vessel. In such circumstances anaerobic stabilization of the major part of the organic matter may be a suitable method of treatment in spite of its lower efficiency and slow rate of reaction. The basic difference between aerobic and anaerobic oxidation is that in the aerobic system oxygen is the ultimate hydrogen acceptor with a large release of energy. In the anaerobic system the ultimate hydrogen acceptor may be nitrate, sulphate or various organic compounds, resulting in a much lower release of energy. Complete stabilization of organic matter cannot be achieved anaerobically, and it is normally necessary to treat the anaerobic plant effluent further by aerobic means if it is to be discharged directly to a receiving water.

As seen in Figure 6.1, anaerobic oxidation is a two-stage process and as a result has certain operational problems. The acid-forming bacteria which carry out the first stage of the breakdown are fairly adaptable as regards environmental conditions, but the methane-formers responsible for the second stage are more sensitive. In particular the methane-formers will only operate effectively in the pH range 6.5–7.5. It is thus important to control conditions to suit the methane bacteria. Overproduction of acids by the fast-acting acid-formers can rapidly result in a low pH, thus stopping the action of the methane-formers and leaving the reaction at a point where particularly unpleasant and odoriferous compounds have been produced. Further production of acid will lower the pH to such a level that even the acid-formers are inhibited and all action will cease. Matters can then only be rectified by pH correction with chemicals, usually lime, but prevention of such mishaps is a better solution and is achieved by careful observation of pH and volatile acids concentration. Both types of bacteria prefer warm conditions and optimum temperatures for anaerobic oxidation are about 35°C or 55°C.

Reference

Thomas, H. A. (1950). Graphical determination of BOD rate constants. *Wat. Sew. Wks*, **97**, 123.

Further reading

Aziz, J. A. and Tebbutt, T. H. Y. (1980). Significance of COD, BOD and TOC correlations in kinetic models of biological oxidations. *Wat. Res.*, **14**, 319.

Horan, N. J. (1989). *Biological Wastewater Treatment Systems*. Chichester: Wiley.

McKinney, R. E. (1962). *Microbiology for Sanitary Engineers*. New York: McGraw-Hill.

Simpson, J. R. (1960). Some aspects of the biochemistry of aerobic organic waste treatment, *and* Some aspects of the biochemistry of anaerobic digestion. In *Waste Treatment* (P. C. G. Isaac, ed.). Oxford: Pergamon.

Tebbutt, T. H. Y. and Berkun, M. (1976). Respirometric determination of BOD. *Wat. Res.*, **10**, 613.

Tyers, R. G. and Shaw, R. (1989). Refinements to the BOD test. *J. Instn Wat. Envir. Managt*, **3**, 366.

Problems

1. The analysis of a wastewater is

Organic carbon	325 mg/l
Organic nitrogen	50 mg/l
Ammonia nitrogen	75 mg/l
Nitrite nitrogen	5 mg/l

   Calculate the ultimate oxygen demand. (1445 mg/l)

2. Laboratory determinations on an industrial waste indicate that its ultimate BOD is 750 mg/l and the k value at 20°C is 0.20/day. Calculate the five day BOD. What would be the five day BOD if the k value dropped to 0.1/day? (675 mg/l, 510 mg/l)

3. Three samples all have the same five day BOD of 200 mg/l, but their k values are 0.10, 0.15 and 0.25/day. Determine the ultimate BOD for each sample. (295 mg/l, 244 mg/l, 212 mg/l)

4. A domestic sewage has a five day BOD at 20°C of 240 mg/l. If the k value is 0.1/day, determine the BOD at one and five days at 13°C. (46.5 mg/l, 171 mg/l)

5. A series of BOD (ATU) determinations was made in a sample to enable calculation of the ultimate BOD and rate constant. Incubation was carried out on a 5 per cent dilution of the sample at 20°C when the initial saturation DO for samples and blanks was 9.10 mg/l.

Day	Final DO in sample (mg/l)	Final DO in dilution water (mg/l)
1	7.10	9.00
2	6.10	9.00
3	5.10	8.90
4	4.20	8.90
5	3.90	8.80
6	3.50	8.70
7	3.00	8.60

   Use the Thomas method to calculate the L and k values for the sample. (126 mg/l, 0.145/day)

7

Water pollution and its control

It is important to appreciate that all natural waters contain a variety of contaminants arising from erosion, leaching and weathering processes. To this natural contamination is added that arising from domestic and industrial wastewaters which may be disposed of in various ways, e.g. into the sea, on to land, into underground strata or, most commonly, into surface waters.

Any body of water is capable of assimilating a certain amount of pollution without serious effects because of the dilution and self-purification factors which are present. If additional pollution occurs the nature of the receiving water will be altered and its suitability for various uses may be impaired. An understanding of the effects of pollution and of the control measures which are available is thus of considerable importance to the efficient management of water resources.

7.1 Types of pollutant

Contaminants behave in different ways when added to water. Non-conservative materials including most organics, some inorganics and many microorganisms are degraded by natural self-purification processes so that their concentrations reduce with time. The rate of decay of these materials is a function of the particular pollutant, the receiving water quality, temperature and other environmental factors. Many inorganic substances are not affected by natural processes so that these conservative pollutants can only have their concentrations reduced by dilution. Conservative pollutants are often unaffected by normal water and wastewater treatment processes so that their presence in a particular water source may limit its use.

As well as the classification into conservative or non-conservative characteristics the following constituents of pollutants are of importance.

1. Toxic compounds which result in the inhibition or destruction of biological activity in the water. Most of these materials originate from industrial discharges and would include heavy metals from metal finishing and plating operations, moth repellents from textile manufacture, herbicides and pesticides, etc. Some species of algae can release potent toxins and cases have

been recorded where cattle have died after drinking water containing algal toxins.
2. Anything which may affect the oxygen balance of the water, including
    (a) substances which consume oxygen: these may be organic materials which are biochemically oxidized or inorganic reducing agents;
    (b) substances which hinder oxygen transfer across the air–water interface. Oils and detergents can form protective films at the interface which reduce the rate of oxygen transfer and may thus amplify the effects of oxygen-consuming substances;
    (c) thermal pollution, which can upset the oxygen balance because the saturation DO concentration reduces with increasing temperature.
3. Inert suspended or dissolved solids in high concentrations can cause problems, e.g. china-clay washings can blanket the bed of a stream preventing the growth of fish food and removing fish from the vicinity as effectively as a direct poison. The discharge of saline mine drainage water may render a river unsuitable for water-supply purposes.

It is obviously important to be able to assess the effect of a particular polluting discharge on a receiving water in quantitative terms and a first step is to utilize a mass balance approach. Figure 7.1 shows a river receiving a pollutant discharge

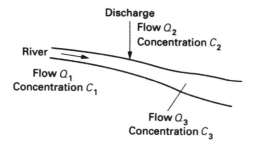

Figure 7.1 The mass balance concept in river pollution.

and it is possible to determine the downstream concentration of the pollutant, assuming instantaneous mixing with conservation of mass

$$Q_1 \times C_1 + Q_2 \times C_2 = Q_3 \times C_3 \qquad (7.1)$$

Since the sum of the flows arriving and leaving the discharge point must be equal (i.e. $Q_3 = Q_1 + Q_2$) the downstream concentration C_3 is easily calculated. Depending upon the nature of the pollutant it will then be possible to calculate the concentrations at points further downstream from the discharge, knowing the velocity of flow and hence the time of travel between the points.

Worked example on mass balance

A stream with flow of $0.1\,\text{m}^3/\text{s}$ and chloride concentration of $52\,\text{mg/l}$ receives a discharge of mine drainage water with a flow of $0.025\,\text{m}^3/\text{s}$ and a chloride concentration of $1250\,\text{mg/l}$.

By mass balance, $0.1 \times 52 + 0.025 \times 1250 = (0.1 + 0.025) \times$ downstream concentration; hence, downstream concentration $= (5.2 + 31.25)/0.125 = 291.6\,\text{mg/l}$.

Chloride is a conservative pollutant so that the concentration will only be reduced below $291.6\,\text{mg/l}$ if additional water with a lower chloride concentration enters the stream below the drainage discharge. In the case of non-conservative pollutants the initial concentration will decrease downstream due to decay reactions.

7.2 Self-purification

In a natural water, self-purification exists in the form of a biological cycle (Figure 7.2) which is able to adjust itself, within limits, to changes in the environmental conditions. In a low organic-content stream there is little nutrient material to

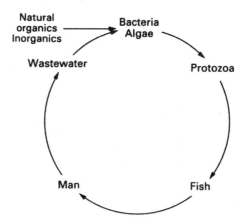

Figure 7.2 The self-purification cycle.

support life so that although many different types of organisms may be present there are only relatively low numbers of each type. In streams with high organic content it is likely that the DO level will be severely depressed, producing conditions unsuitable for animals and higher plant life. In these circumstances bacteria will predominate, although given sufficient time the organic matter will be stabilized, the oxygen demand will fall and a full range of life forms will appear again.

Self-purification involves one or more of the following processes.

- Sedimentation, possibly assisted by biological or mechanical flocculation. The deposited solids will form benthal deposits which, if organic, will decay anaerobically and which, if resuspended by flood flows, can exert sudden high oxygen demands on the system.
- Chemical oxidation of reducing agents such as sulphides.
- Bacterial decay due to the generally inhospitable environment for enteric and pathogenic bacteria in natural waters.
- Biochemical oxidation which is normally by far the most important process. To prevent serious pollution it is important that aerobic conditions are maintained; this means that the balance between oxygen consumed by BOD and that supplied by reaeration from the atmosphere is not drastically disturbed.

Reaeration

In the absence of any external mixing the concentration of a gas dissolved in water will eventually become uniform due to molecular diffusion. The rate of diffusion is proportional to the concentration gradient and is described by Fick's law

$$\frac{\partial M}{\partial t} = k_{\mathrm{d}} A \frac{\partial C}{\partial l} \tag{7.2}$$

where M = mass transfer in time t,
$\quad\quad k_{\mathrm{d}}$ = diffusion coefficient,
$\quad\quad A$ = cross-sectional area across which transfer occurs,
$\quad\quad C$ = concentration and
$\quad\quad l$ = distance in direction of transfer.

A solution to equation 7.2 is

$$C_t = C_{\mathrm{s}} - 0.811(C_{\mathrm{s}} - C_0) \left(\mathrm{e}^{-K_{\mathrm{d}}} + \frac{1}{9} \mathrm{e}^{-9K_{\mathrm{d}}} + \frac{1}{25} \mathrm{e}^{-25K_{\mathrm{d}}} + \ldots \right) \tag{7.3}$$

where C_0 = concentration at time 0,
$\quad\quad C_t$ = concentration at time t,
$\quad\quad C_{\mathrm{s}}$ = saturation concentration and
$\quad\quad K_{\mathrm{d}}$ = $k_{\mathrm{d}} \pi^2 t / 4 l^2$

The diffusion coefficient (k_{d}) is usually expressed as $\mathrm{mm^2/s}$ and for oxygen in water has a value of $1.86 \times 10^{-3}\,\mathrm{mm^2/s}$ at 20°C.

The solution of a gas in liquid is governed by two physical laws, Dalton's law of partial pressures and Henry's law.

Dalton's law of partial pressures states that the partial pressure of a gas in a mixture of gases is the product of the proportion of that gas in the mixture and the total pressure.

Henry's law states that at constant temperature, the solubility of a gas in a liquid is proportional to the partial pressure of the gas.

The rate of solution of oxygen is proportional to the saturation deficit, D, i.e.

$$\frac{dD}{dt} = -KD \tag{7.4}$$

where K is an aeration constant. Hence

$$D_t = D_a \, e^{-K_2 t} \tag{7.5}$$

where D_t = DO deficit at time t, D_a = DO deficit initially and K_2 = reaeration constant.

Expressing equation 7.5 in terms of DO concentrations, we get

$$\log_e \frac{(C_s - C_t)}{(C_s - C_0)} = -K_2 t \tag{7.6}$$

The reaeration constant is a function of the velocity of flow, channel configuration and temperature. A number of empirical relationships are used for prediction of the value of K_2 and most depend upon velocity and depth of flow for a given temperature as indicated below

$$K_2 = \frac{cv^n}{h^m} \tag{7.7}$$

where v = mean velocity (m/s), h = mean depth (m) and c, n and m are constants with typical values of 2.1, 0.9 and 1.7 respectively.

Rather than using the reaeration constant K_2 it is sometimes preferable to adopt a parameter which measures the reaeration rate per unit area exposed per unit DO deficit, usually termed the exchange coefficient f, given by

$$f = K_2 \frac{V}{A} \tag{7.8}$$

where V = volume of water below interface and A = area of air–water interface. $K_2(V/A)$ is termed the aeration depth and has units of velocity. Table 7.1 gives

Table 7.1 Typical values of the exchange coefficient f

Situation	f (mm/h)
Stagnant water	4–6
Water in channel at 0.6 m/min	10
Sluggish polluted river	20
Thames estuary	55
Water in channel at 10 m/min	75
Open sea	130
Water in channel at 15 m/min	300
Turbulent Lakeland beck	300–2000
Water flowing down 30° slope	700–3000

After Klein (1962).

typical values of the exchange coefficient. Studies on a number of British rivers (Owens *et al.* 1964) have shown that the value of f in mm/h at 20°C can be predicted from the formula

$$f = 7.82 \times 10^4 U^{0.67} H^{-0.85} \tag{7.9}$$

where U = velocity of water (m/s) and H = mean depth of flow (mm).

A rise in temperature of 1 degree increases the value of f by about 2 per cent and similarly a fall in temperature decreases the rate of reaeration.

One of the problems in the study of rivers is to determine the reaeration characteristics of a stream. Solution of oxygen can only occur at the air–water interface where a thin film of water is rapidly saturated and further reaeration is then controlled by the diffusion of oxygen throughout the main body of water which is a slow process. In a turbulent stream this saturated surface layer is broken up and reaeration can proceed more rapidly (Figure 7.3).

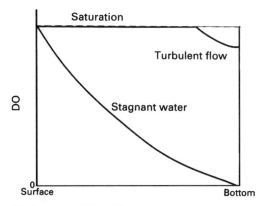

Figure 7.3 Effect of turbulence on DO profiles.

Field determination of the reaeration characteristics of a stream (Gameson *et al.* 1955) involves partial deoxygenation of the stream with a reducing agent (sodium sulphite plus cobalt catalyst) and measuring DO uptake at stations downstream. Assuming there is no significant BOD or photosynthesis in the reach,

$$f = \frac{V}{t} \frac{1}{A} \log_e \frac{(C_s - C_1)}{(C_s - C_2)} \tag{7.10}$$

where C_1 and C_2 are DO concentrations at two stations downstream of the reagent addition with a time of flow between them of t where $V/t = Q$, the rate of flow.

The sag curve

The situation which occurs in a stream receiving a single pollution load is shown in Figure 7.4. If the stream is originally saturated with DO the BOD uptake curve for the mixture of effluent and stream water gives the cumulative deoxygenation of the stream. As soon as BOD begins to be exerted the DO falls below saturation and reaeration starts. With increasing saturation deficit the rate of reaeration increases until a critical point is reached where the rates of deoxygenation and reaeration are equal. At the critical point, minimum DO is reached and as further time passes the DO will increase.

Assuming that the only processes involved are BOD removal by biological oxidation and DO replenishment by reaeration from the atmosphere, the Streeter–Phelps equation was derived

$$\frac{dD}{dt} = K_1 L - K_2 D \tag{7.11}$$

where D = DO deficit at time t,
L = ultimate BOD,
K_1 = BOD reaction rate constant and
K_2 = reaeration constant.

Integrating and changing to base 10 ($k = 0.4343\,K$), we get

$$D_t = \frac{k_1 L_a}{k_2 - k_1} (10^{-k_1 t} - 10^{-k_2 t}) + D_a 10^{-k_2 t} \tag{7.12}$$

where D_a and L_a are values at $t = 0$ and the ultimate BOD remaining at time t is $L_t = L_a 10^{-k_1 t}$.

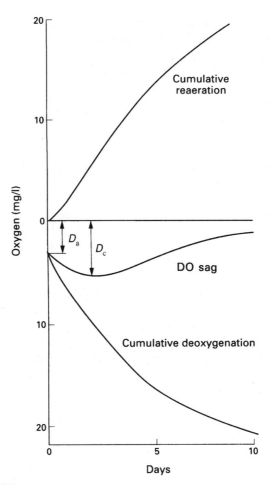

Figure 7.4 The DO sag curve.

The critical point, i.e. the point of maximum deficit, is given by

$$\frac{\mathrm{d}D}{\mathrm{d}t} = 0 = K_1L - K_2D \tag{7.13}$$

Hence

$$D_c = \frac{k_1}{k_2} L_a 10^{-k_1 t_c} \tag{7.14}$$

$$t_c = \frac{1}{k_2 - k_1} \log \frac{k_2}{k_1} \left[1 - \frac{D_a(k_2 - k_1)}{L_a k_1} \right] \tag{7.15}$$

where D_c = critical deficit which occurs at time t_c.

This equation may be used in a number of ways although in theory it is only valid when there is no change in dilution or pollution load in the stretch under consideration.

For more complex river systems, a step-wise calculation can be adopted, treating each section of river between changes in pollution and/or flow as an individual problem. The calculated output BOD and DO from this section then provide part of the input data for the next section downstream. The calculation process can be repeated as often as required to make a predictive mathematical model for a particular river system. Such a model can also include other non-conservative pollutants using theoretical or empirical decay relationships analogous to the Streeter–Phelps expression for DO.

A variety of factors may come into play in the oxygen balance, probably the most important being deposition of organic matter from suspension and possible later resuspension due to scour of bottom mud. The contribution of bottom muds to oxygen demand can be considerable in shallow wide channels but relatively small in narrow deep channels. Flood flows can resuspend muds, giving a very high oxygen demand.

Other factors which may influence the DO sag include

- BOD addition in surface runoff
- removal of DO by diffusion into bottom mud to satisfy oxygen demand
- BOD addition by diffusion of soluble organics from bottom deposits
- removal of DO by purging action of gases released from bottom deposits
- addition of DO by photosynthetic activities of plants
- removal of DO by plants during night
- continuous redistribution of DO and BOD by longitudinal dispersion.

Dobbins (1964) produced a method of allowing, in part at least, for the above factors. It was found that longitudinal dispersion of BOD and DO in most freshwater streams was negligible. Accurate measurement of surface reaeration was most important.

Worked example for dissolved oxygen sag

A stream with a flow of $0.3 \, \text{m}^3/\text{s}$, DO 7 mg/l and BOD 2 mg/l receives an effluent discharge of $0.2 \, \text{m}^3/\text{s}$, DO 1.2 mg/l and BOD 22 mg/l. The BOD rate constant (k_1) for the stream is 0.1/day and the reaeration constant (k_2) is 0.35/day. Calculate the DO concentration at a point three days time of travel below the effluent discharge. Saturation DO is 9.1 mg/l.

Initial DO $= [(0.3 \times 7) + (0.2 \times 1.2)]/(0.3 + 0.2) = 4.68 \, \text{mg/l}$
Initial DO deficit $= 9.1 - 4.68 = 4.42 \, \text{mg/l}$
Initial BOD $= [(0.3 \times 2) + (0.2 \times 22)]/(0.3 + 0.2) = 10.0 \, \text{mg/l}$

This initial five-day BOD needs to be converted into the ultimate BOD using equation 6.12

$$\text{Initial ultimate BOD} = 10/(1 - 10^{-0.1 \times 5}) = 14.60 \, \text{mg/l}$$

Now using equation 7.12 the DO deficit after three days will be

$$D_3 = [(0.1 \times 14.60)/(0.35 - 0.1)](10^{-0.1 \times 3} - 10^{-0.35 \times 3}) + (4.42 \times 10^{-0.35 \times 3})$$
$$= 2.8 \, \text{mg/l}$$

Hence the DO concentration after three days is $9.1 - 2.8 = 6.3 \, \text{mg/l}$.

7.3 Toxic materials

Fish are usually used as sensitive indicators of toxic pollution but the situation is complicated because various environmental factors can considerably affect the toxicity of a particular material. As far as fish are concerned the two most important environmental factors are DO and temperature. Fish require a certain minimum oxygen supply for normal activity, ranging from about 1.5 mg/l for certain coarse fish to 5 mg/l for game fish. At or near these limiting DO levels the activity of fish may be impaired so that their sensitivity to poisonous materials is often increased. Certain poisons such as the heavy metals interfere with respiration so that their harmful properties are enhanced at low DO. Thus a reduction in DO to 50 per cent saturation will reduce the concentration at which a heavy metal becomes toxic to about 70 per cent of the concentration which is toxic in oxygen-saturated water.

The metabolic rate of fish is closely linked with temperature so that a rise of 10°C will increase the oxygen requirement by two or three times. Unfortunately the saturation concentration of DO falls with increasing temperature so that the effect of raising the temperature is to raise the oxygen requirement whilst simultaneously reducing the available oxygen supply. As a rough guide it can be taken that a 10°C rise in temperature will approximately halve the concentration at which material is toxic.

Another factor which can have a considerable effect on toxicity is pH, a good example of this being found in the behaviour of ammonium compounds which are relatively innocuous at low pH values. Under alkaline conditions, however, ammonia can be quite harmful to fish and a rise in pH from 7.4 to 8.0 can halve the toxic concentration. It appears that unionized ammonia is the toxic form, ionized ammonia, which predominates at low pH values, being much less toxic. In general unionized substances are more readily absorbed by fish than the ionized forms.

The presence of dissolved salts in water is a further factor which can influence the toxicity of certain substances. The presence of calcium ions in solution will considerably reduce the toxic effect of heavy metals such as lead and zinc. High concentrations of sodium, calcium and magnesium prevent the toxic effects of heavy metals probably by forming complexes with them. For example, 1 mg/l of lead in a soft water may be rapidly fatal to fish, but in a hard water of, say, 150 mg/l calcium hardness, 1 mg/l of lead will not be harmful.

The effect of potentially toxic materials in rivers has often been measured by their action on fish as demonstrated by some form of bioassay. The procedure involves the use of a series of dilutions of the suspect material to which test fish are exposed under standard conditions. The prescribed measure of acute toxicity is the median tolerance limit (TL_m) sometimes referred to as the 50 per cent lethal dose (LD_{50}). This is the concentration of material under test at which 50 per cent of the test fish are able to survive for a specified period of exposure (usually 48 or 96 h). Table 7.2 gives some typical values of toxic levels as a guide although because of the many variations in procedure and in environmental conditions it is not possible to state that fish can only tolerate a certain concentration of a particular material. The number of fish and time required for chronic toxicity testing can be a considerable disadvantage so that other techniques are becoming popular. Procedures using the crustacean *Daphnia*

Table 7.2 Some compounds toxic to fish

Material	*Occurrence*	*Approx. LD_{50} mg/l*
Acridine	Coal-tar wastes	0.7–1.0
Aldrin	Insecticide	0.02
Alkyl benzene sulphonate	Sewage effluent	3–12
Ammonia	Sewage effluent	2–3
Chloramine	Chlorinated effluents	0.06
Chlorine	Chlorinated effluents	0.05–0.2
Copper sulphate	Metal processing	0.1–2.0
	Algal control of reservoirs	
Cyanide	Plating wastes	0.04–0.1
DDT	Insecticide	<0.1
Detergents, synthetic (packaged)	Sewage effluent	15–80
Fluoride	Aluminium smelting	2.5–6.0
Gammexane	Insecticide	0.035
Hydrogen sulphide	Bottom muds, sludge	0.5–1.0
Methyl mercaptan	Oil refineries	1.0
	Wood pulp processing	
Naphthalene	Coal-tar wastes	10–20
	Gas liquor	
Parathion	Insecticide	0.2
Potassium dichromate	Flow gauging	50–500
Silver nitrate	Photographic wastes	0.004
Zinc	Galvanizing	1–2
	Rayon manufacture	

Note: These figures are intended only as a guide, the actual LD_{50} in any particular situation will depend on environmental factors, the species of fish involved and the duration of the exposure.

magna and an alga, *Selenastrum*, provide alternative toxicity tests which are fairly widely used by pollution control organizations.

Commercial toxicity testing instruments are now available which rely on the toxic substance preventing or inhibiting the activity of a chemiluminescent enzyme or of a bioluminescent bacterium added to the sample. The luminescence of treated samples is measured on a calibrated instrument so that a quantitative measurement of the degree of inhibition, and thus the toxic effect, can be obtained.

With all toxicity testing the procedures involve exposure of the test organism or enzyme to serial dilutions of the suspect substance with the objective of determining the 'no observed effect concentration' (NOEC). It is important to understand that no particular type of toxicity test is more sensitive to all toxic substances so in some cases it may be necessary to employ a battery of tests to obtain the required data.

When considering raw waters for potable supply the presence of toxic substances must always be seen as a potential hazard. The use of lowland rivers for potable supply inevitably implies increased potential danger because of accidental discharges of toxic materials. Unfortunately it is not feasible to analyse a raw water for all known toxic compounds and reliance must largely be placed on rapid reporting of accidental discharges of such compounds. The recommended seven day bankside storage of raw water from lowland rivers provides a degree of safeguard, but there is nevertheless a need for some form of monitoring device to warn of the presence of toxic material. A universal monitoring device is not likely to be available, but a number of techniques using fish or microorganisms offer at least some degree of warning of acute levels of toxic material. The question of monitoring for trace levels of toxic compounds is much more difficult.

With some industrial discharges there may be concerns that a complex cocktail of substances could be harmful to aquatic life in the receiving water. Such effects could occur at concentrations of pollutants which might be below the limits of detectability for oxygen demand or they might be caused by inorganic substances which exert no oxygen demand and contain no organic carbon. In protecting the water environment it would be very difficult and time consuming to analyse for all possible toxic constituents, particularly since their presence may vary depending upon the stage of an industrial process at which samples are taken.

7.4 Overall effects of pollution

When considering pollution by wastewaters there are of course effects other than the creation of DO deficits. Depending on the dilution available there may be significant increases in dissolved solids, organic content, nutrients such as nitrogen and phosphorus, colour and turbidity. All of these constituents may give rise to undesirable changes in water quality particularly as regards downstream abstraction. Nutrient build-up can be a serious problem in lakes and very slow-moving waters but is not likely to be so troublesome with rivers. It should be remembered, however, that in many river abstraction schemes for water supply, raw water is stored in large shallow reservoirs prior to treatment. Even quite low nutrient contents in the water can result in prolific seasonal algal growths, sometimes making the water much more difficult to treat than the original river water.

All lakes undergo a natural change in their characteristics which, in the absence of human activity, may take thousands of years. A lake in a 'wilderness' catchment receives inflow from largely barren surroundings and thus collects little in the way of organic food and inorganic nutrients. Such nutrient-deficient waters are termed oligotrophic and are characterized by low TDS levels, very low turbidities and small biological populations. As the catchment becomes older (in almost geological time scale) there is a gradual increase in nutrient levels and hence in biological productivity, with a consequent deterioration in water quality. Eventually, as nutrient levels and biological production increase, the water becomes nutrient-rich or eutrophic. Nutrients are recycled and in extreme cases the water may become heavily polluted by vegetation, low DO levels will occur due to rotting plants and during darkness, anaerobic conditions may well exist. The eventual fate of all lakes is to become eutrophic, but the rate at which this end point is reached can be greatly accelerated by artificial enrichment due to human activities. Nitrogen and phosphorus are the most important nutrients in the context of eutrophication and since some algae can fix atmospheric nitrogen it is generally accepted that phosphorus is the limiting nutrient in water. The level of phosphorus above which algal growth becomes excessive depends upon many factors, but in UK conditions waters with a winter phosphate level less than 5 μg/l are unlikely to exhibit eutrophic tendencies. Phosphates occur in sewage effluents due partly to human excretion and partly to their use in synthetic detergents.

Serious pollution which often occurs in industrialized areas can have very profound effects on a river system and reduction of water pollution in such a system is inevitably an expensive operation, usually taking many years to achieve. Ideally it would be desirable for every river to be unpolluted, full of fish and aesthetically pleasing. In an industrialized country it becomes economically impossible to prevent all river pollution and it is necessary to take an overall view of water resources and to classify rivers as suitable for particular purposes.

River pollution is clearly undesirable for many reasons

- contamination of water supplies – additional load on treatment plants
- restriction of recreational use
- effect on fish life
- creation of nuisances – appearance and odour
- hindrance to navigation by banks of deposited solids.

A typical water use classification might thus be (in decreasing order of quality requirements)

1. domestic water supply
2. industrial water supply
3. commercial fishing
4. irrigation
5. recreation and amenity
6. transportation
7. waste disposal.

Each use has specific requirements for quality and quantity of water and some uses may be incompatible. Irrigation is a consumptive use in that water used in this way does not find its way back into the river system. Considerable volumes of cooling water are lost by evaporation. The other uses are not in general consumptive, although they usually have a detrimental effect on quality. Thus water abstracted for domestic supply is returned as sewage effluent. The conservation of water resources depends on multi-purpose use of water wherever possible.

7.5 Groundwater pollution

The straining action of soil and rocks as water percolates through them is normally sufficient to remove suspended impurities from contaminated infiltration flows. It should be noted, however, that excessive suspended solids can accumulate in the pores and thus eventually block the aquifer, preventing further recharge. Soluble impurities may be removed by the ion-exchange properties of some soils and rocks, but this is by no means the case with all contaminants. There is increasing concern in many countries about contamination of aquifers which may seriously inhibit their use as water sources. A major problem in some areas is the presence of high nitrate levels in groundwaters due to increased drainage and heavy fertilizer applications which tend to occur as the result of intensive farming practices. In the EU a nitrate directive requires the designation of 'nitrate vulnerable zones' (NVZs) where the nitrate nitrogen concentration in an aquifer is likely to exceed 50 mg/l. Within a NVZ, action plans are established to control agricultural and other activities which influence nitrate concentrations in the groundwater but it may be many years after the introduction of such measures before significant reductions in groundwater nitrate levels occur.

The use of soakaways for the disposal of domestic and industrial wastewaters and for the removal of surface runoff can produce major groundwater quality problems and there are potential hazards from fuel storage installations and oil pipelines. The seepage from solid wastes tips can be highly polluting and strict planning controls are enforced in many areas to prevent construction of such facilities where groundwater pollution is possible.

Organic matter entering an aquifer will be stabilized very slowly because the oxygen demand soon deoxygenates the water and produces anaerobic conditions. This can result in the production of unpleasant tasting and smelling compounds as well as causing the dissolution of iron from the surrounding rocks. A major problem with groundwater pollution is the lack of significant self-purification capacity so that once polluted, an aquifer may become useless for water supply purposes for the foreseeable future. In the UK about 30 per cent of public water supply is derived from groundwater, in the USA about 50 per cent and in Denmark 99 per cent. Protection of groundwater quality is thus a vital aspect of water quality control and conservation of water resources. Particular care must be taken to protect important aquifers and in some cases underground disposal of liquid wastes and solid waste tips with leachate problems may only be permitted if the aquifer is known to be completely isolated from the potential source of pollution. A groundwater protection strategy to protect aquifers can be developed based on an assessment of the vulnerability of individual aquifers to contamination. Protection zones can then be established to ensure that potentially polluting activities in the catchment are closely controlled and the risks of contamination minimized. The groundwater protection policy used in England and Wales identifies zones around a groundwater abstraction

- zone I (inner source protection) – immediately adjacent to the groundwater abstraction and defined as an area within a 50 day travel time of the abstraction
- zone II (outer source protection) – area within a 400 day travel time of the abstraction
- zone III (source catchment) – complete catchment of the source.

Landfills of any type are prohibited in zone I; domestic, inert and construction waste landfills are permitted in zone II with adequate operational safeguards; highly polluting industrial waste landfills would only be permitted in zone III if they had an engineered containment system. Similar constraints apply to the land disposal of liquid wastes, sludges and slurries, surfacewater drainage and wastewater effluents in the various zones.

7.6 Pollution of tidal waters

For many years communities with access to tidal waters have utilized such waters as a convenient disposal facility. The potential for dilution and dispersion of pollutants in the open sea is considerable and there is a large self-purification capacity. This does not mean that the seas can be considered as an infinite sink for the disposal of unwanted materials nor does it mean that all tidal waters are suitable for sewage discharges. The upper reaches of a tidal estuary are likely to have pollution-assimilating characteristics similar to those of the non-tidal reaches of the river. Narrow tidal estuaries or those with land-locked exits often have complex flow characteristics which mean that discharges can take several days to travel a relatively short distance to the open sea. On the other hand, sewage discharges to deep open water with strong tidal flows are likely to be almost undetectable. There is strong pressure from environmental groups to end the discharge of untreated sewage to all tidal waters and such disposal practices will be illegal in the EU within a few years. The scientific validity of a blanket ban on the discharge of untreated sewage to all tidal waters is dubious at best, although it can be justified in certain circumstances. Provided discharges to the open sea are made through long outfalls with well designed diffusers on the sea bed any environmental effects are likely to be negligible. When discharges are made to areas near bathing beaches and shellfish beds or where ambient temperatures are high and tidal action is limited it may be necessary to install full conventional sewage treatment as used for inland discharges. It may also be desirable to disinfect the effluent before release when bathing waters are involved. Eutrophication can occur in land-locked estuaries and bays with prolific growths of algae creating a public nuisance so that in some situations nutrient control in the catchment area may be necessary. Problems can also arise in coastal waters where nutrient discharges can encourage the growth of dinoflagellate algae, some of which release powerful toxins which can be dangerous to consumers of shellfish feeding in the area.

7.7 Control of pollution

Because of the need to reconcile the various demands on the aquatic environment and on water resources most countries have pollution control bodies to maintain and hopefully improve water quality. It is perhaps worth quoting at this point the European Commission statement that water pollution means

> . . . the discharge by man of substances into the aquatic environment the results of which are such as to cause hazards to human health, harm to living resources and aquatic ecosystems, damage to amenities or interfere with other legitimate uses of water.

It follows that for a discharge to be termed polluting there must be evidence of actual harm or damage.

Water quality management

As an example of modern concepts in water management it is useful to consider how in England and Wales all aspects of the hydrological cycle came under the control of the National Rivers Authority (NRA) in 1989. The Authority operated on a regional basis with responsibility within the catchment of a major river or rivers for monitoring of water quality, control of pollution, management of water resources for public supply, provision of effective flood defences, improvement and development of fisheries, conservation and protection of the water environment and the promotion of water-based recreational activities. The NRA granted consents to discharge effluents to surfacewaters and groundwaters subject to the discharge meeting appropriate standards as to composition and flow. If certain particularly hazardous ('red list') substances, shown in Table 7.3 were present, consents to discharge were set by Her Majesty's Inspectorate of Pollution (HMIP) in consultation with the NRA. The discharge of these prescribed substances must be prevented or minimized to render harmless any release.

Table 7.3   UK 'red list' dangerous substances

Mercury and its compounds
Cadmium and its compounds
Lindane
DDT
Pentachlorophenol (PCP)
Hexachlorobenzene (HCB)
Hexachlorobutadiene (HCBD)
Aldrin
Dieldrin
Endrin
Polychlorinated biphenyls (PCB)
Tributyltin compounds
Triphenyltin compounds
Dichlorvos
Trifluralin
1,2-Dichloroethane
Trichlorobenzene
Azinphos-methyl
Fenitrothion
Malathion
Endosulfan
Altrazine
Simazine

This was part of the policy of 'integrated pollution control' (IPC) which aims to ensure that such substances do not contaminate any sector of the environment. Similar mechanisms are used in most developed countries by either regional or national regulatory authorities. Increasing public awareness of the water environment has caused more emphasis to be placed on water pollution control activities and the adoption of 'best available technology' (BAT) treatment processes, particularly for industrial wastewaters, may be stipulated. As technologies develop it may become possible to achieve higher effluent qualities and the BAT concept may therefore contribute to a 'ratchet' effect on consent standards. BAT may involve significant extra costs over previous treatment techniques and this is recognized by the use of the term BATNEEC (best available technology not entailing excessive cost). As effluent standards become more demanding the financial implications for the public and industry may reach the point where the economic solution may be to cease the discharge. This is not really practical for domestic wastewater discharges but could sometimes be an option for industry, resulting in a change of process, or closure of the production facility. The latter approach can of course have major political implications if, as is likely, it results in job losses. At that stage it becomes necessary to try to balance environmental benefits against unemployment in a highly emotive arena.

In 1996 a new Environment Agency was established in England and Wales taking over the duties and responsibilities of the NRA, HMIP and the waste regulation authorities which were previously part of local government. The new Agency provides a fully integrated approach to environmental management which should ensure that quality is maintained and enhanced in the air, land and water sectors of the environment. Local Environment Agency Plans (LEAPs) aim to arrive at agreed solutions to environmental management and build on the experience gained by the NRA in its preparation of catchment management plans (CMPs). The Scottish Environment Protection Agency (SEPA) has a similar role in Scotland.

The process of catchment management planning allows an independent regulator to balance competing requirements and the differing interests of all users of a river system. The environmental potential of a catchment can then be realized in terms of water quality, water quantity and physical features. A catchment management plan focuses on a river and its associated corridor to analyse the issues which affect the catchment and to suggest solutions to resolve problems and conflicts. Many of the issues in a catchment can only be addressed with the co-operation and assistance of other bodies, organizations or industries, so that the preparation of a CMP must involve consultation with local communities and others with an interest in the matter.

Standards for water pollution control can be based either on the quality required in the receiving water (the 'water quality objective' or WQO approach) or they can be applied directly to the discharge without reference to the conditions relating to the receiving water (the 'emission standard' approach). The WQO approach is more logical since it can take into account the dilution available and the other uses of the receiving water. It can, however, cause problems if a new discharge is made to the system since all existing discharge consents must be revised downward or the new discharge may be required to attain an impossibly high standard. There could be inequalities in the degree of treatment required for similar wastewaters discharged to different reaches of the river because of variations in the assimilative capacity. Emission standards are administratively convenient in that they are applied across the board to all similar discharges. However, because they make no allowance for the specific characteristics of a particular location, such as self-purification capacity and downstream water use, emission standards are often economically and scientifically unsound. In practice a combination of fixed emission standards and the WQO approach may offer some practical and economic advantages. Whatever method is used for the control of water pollution the primary objectives are to safeguard public and industrial water abstractions, to safeguard public health, to maintain and improve fisheries, to maintain and restore water quality and to conserve aquatic flora and fauna.

Water pollution control in the UK was for many years largely based on the pioneering work of the Royal Commission on Sewage Disposal which in its Eighth Report (1912) proposed the adoption of effluent standards related to the conditions prevailing in the receiving water. In its studies the Commission concluded that a BOD of 4 mg/l in a watercourse was a limit of acceptability which, if exceeded, would indicate a significant degree of pollution. A clean river was believed to have a BOD of 2 mg/l and when considering a typical sewage works effluent BOD of 20 mg/l it can be calculated that a dilution of 8:1 with clean river water is necessary to prevent the downstream BOD exceeding 4 mg/l. This concept gave rise to the Royal Commission effluent standard of 20 mg/l BOD and 30 mg/l SS which was taken as applying as a 75 percentile value. In many cases effluents were discharged at the 20:30 standard but, because of insufficient dilution, pollution could be serious. It must be said, however, that the Commission was somewhat pessimistic in its choice of 4 mg/l BOD as a limit. There are in fact many watercourses which have the oxygenation capacity to assimilate considerably more than 4 mg/l BOD without significant environmental damage.

A major research programme on urban pollution management (UPM), with particular emphasis on urban rivers, funded by government, industry and the research councils has been undertaken in the UK to aid in the development of pollution control strategies. This project has involved study of the performance of stormwater overflows, the influence of storm profiles on system performance and the contributions of suspended matter to the pollution load. In urban areas sewage discharges from combined sewer overflows (CSOs) can be a major cause of river pollution, particularly where the sewerage systems are in need of rehabilitation. In order to upgrade sewerage systems effectively it is necessary to quantify the effect of the storm overflows on receiving water quality. A suite of computer models permits simulation of a particular river system and enables rational formulation of discharge consents and control policies as well as providing valuable information on optimizing investment in the system for the best environmental improvements.

Water quality classifications

In England and Wales a policy of local emission standards has been used to set discharge consents, originally in relation to a water quality classification scheme developed by the National Water Council (NWC) in 1970 and shown in Table 7.4.

Table 7.4   The NWC river water quality classification

River class	Quality criteria	Remarks	Potential uses
1A	DO >80% saturation BOD not >3 mg/l Ammonia not >0.4 mg/l Complies with A2 water Non-toxic to fish*	Average BOD not >1.5 mg/l No visible evidence of pollution	Potable supply Game fishery High amenity
1B	DO >60% saturation BOD not >5 mg/l Ammonia not >0.4 mg/l Complies with A2 water Non-toxic to fish*	Average BOD not >2 mg/l Average ammonia not >0.5 mg/l No visible evidence of pollution	As for 1A
2	DO >40% saturation BOD not >9 mg/l Complies with A3 water Non-toxic to fish*	Average BOD not >5 mg/l May be some colour and foam	Potable supply after advanced treatment Coarse fishery Moderate amenity
3	DO >10% saturation Not anaerobic BOD not >17 mg/l		Low grade supply Polluted to extent that fish are absent
4	DO <10% saturation Anaerobic at times		Grossly polluted Cause nuisance
X	DO >10% saturation		Insignificant waters Object is simply the prevention of nuisance

* European Inland Fisheries Advisory Commission (EIFAC) terms. A2 and A3 refer to treatment requirements under EC directive 76/464/EEC (Table 2.4).

The general aims were to ensure that no deterioration in water quality occurred, the elimination of class 4 waters and the upgrading of class 3 waters to class 2 where possible. The consent conditions for sewage effluents necessary to satisfy these objectives were usually based on BOD, SS and ammonia nitrogen concentrations using 95th percentile values. A problem with the NWC scheme was that in some situations the presence of chemical constituents not included in the classification protocol meant that a water could not support the aquatic life which would be expected from its class. Heavy metals can be toxic to macroorganisms which are used as food by fish but would not appear in the chemical parameters of the classification. In 1993 the National Rivers Authority adopted a revised river quality classification scheme called the GQA (general quality assessment) scheme (Table 7.5) which is still restricted to chemical parameters but uses percentiles to allow for statistical variations in sampling and analysis. This was intended to give a more reliable basis for assessing long term changes in water quality brought about by the implementation of pollution prevention measures.

Table 7.5   GQA chemical grading for rivers and canals

Water quality	Grade	DO (% saturation) (10 percentile)	BOD (ATU) (mg/l) (90 percentile)	Ammonia (mg N/l) (90 percentile)
Good	A	80	2.5	0.25
	B	70	4	0.6
Fair	C	60	6	1.3
	D	50	8	2.5
Poor	E	20	15	9.0
Bad	F	Poorer than E in one or more determinands		

As discussed earlier, there are advantages in controlling pollution by reference to the needs of water uses in the watercourse. This implies the adoption of water quality objectives which relate to specific uses. The NRA therefore developed the concept of 'statutory water quality objectives' (SWQOs) which was somewhat delayed by political and economic considerations with the result that the initial pilot stages were only implemented after the NRA was absorbed into the Environment Agency in April 1996. The purpose of a SWQO is to establish targets, on a statutory basis, which provide an agreed planning framework for regulatory bodies, dischargers, abstractors and river users. When adopted, SWQOs will secure quality improvements to date by providing a legal baseline for existing discharge consents, as well as providing a means for tackling discharges from non-water sectors of industry, agricultural and other diffuse pollution, and the effects of new or revised abstractions.

The SWQO concept requires the development of a suite of water classifications based on various uses such as water supply, irrigation, fisheries and recreation. As a first step, to cover the general 'health' of a watercourse, the 'river ecosystem classification' (Table 7.6) was developed and this formed the basis for the trial SWQOs which were implemented during 1996.

It will be seen that in addition to the chemical parameters used in the NWC and GQA schemes, the river ecosystem scheme includes other parameters which affect aquatic life. It is known that the toxicity of metals like copper and nickel is influenced by other properties of the water environment such as pH and hardness and these effects are included in the classification scheme.

Aquatic life in waters does of course provide a sensitive in-situ monitoring opportunity which can lead to the development of biological classification systems. The need visually to identify and enumerate living organisms in the water under examination makes biological examination time consuming and requires the services of skilled aquatic biologists. Nevertheless, a biological classification scheme based on the data from regular surveys does permit assessment of the overall quality of a water. The biological classification developed in the UK by the NRA and now employed by the Environment Agency

Table 7.6 Standards for river ecosystem classes

Class	DO (% saturation) (10 percentile)	BOD (ATU) (mg/l) (90 percentile)	Ammonia Total (mg/l) (90 percentile)	Ammonia Unionized (mg/l) (95 percentile)	pH (5–95 percentile)	Hardness (CaCO₃) (mg/l)	Dissolved copper (µg/l) (95 percentile)	Total zinc (µg/l) (95 percentile)
RE1	80	2.5	0.25	0.021	6–9	<10 >10 <50 >50 <100 >100	5 22 40 112	30 200 300 500
RE2	70	4.0	0.6	0.021	6–9	<10 >10 <50 >50 <100 >100	5 22 40 112	30 200 300 500
RE3	60	6.0	1.3	0.021	6–9	<10 >10 <50 >50 <100 >100	5 22 40 112	300 700 1000 2000
RE4	50	8.0	2.5	–	6–9	<10 >10 <50 >50 <100 >100	5 22 40 112	300 700 1000 2000
RE5	20	15.0	9.0	–	–	–	–	–

is based on groups (taxa) of macroinvertebrates such as mayfly nymphs, snails, shrimps and worms. These organisms are used because they remain in more or less the same place in a water, have relatively long lives and respond to physical and chemical characteristics in the water. They are thus affected by infrequent pollution events which are often missed by the spot sampling used for GCA monitoring. For the biological assessment eighty-five groups of macro-invertebrates are employed, each including several species with similar tolerances to pollution. Pollution-tolerant groups are given a score of 1 and as pollution tolerance reduces, the group score is increased up to a value of 10 for the least tolerant taxa. The presence of pollution-intolerant taxa at a site suggests better water quality than at a site where only pollution-tolerant taxa are found. The overall assessment of taxa found at a given site are then compared with the taxa which would be expected in a pristine, unpolluted, water of the same type as set out in Table 7.7.

Table 7.7 GQA biological classification scheme

Grade		Outline description
A	Very good	Biology similar or better than expected – high diversity of taxa usually with several species in each – dominance of one taxon rare
B	Good	Biology a little short of expectation – small reductions in pollution-sensitive taxa – moderate increase in individual species of pollution-tolerant taxa
C	Fairly good	Biology worse than expected – many sensitive taxa absent or number of individual species reduced – marked rise in pollution-tolerant species with some present in high numbers
D	Fair	Biology worse than expected – sensitive taxa scarce – pollution-tolerant taxa present with high numbers of some species
E	Poor	Biology restricted to pollution-tolerant species with some taxa dominant in terms of numbers of individual species – sensitive taxa rare or absent
F	Bad	Biology limited to small number of very pollution-tolerant taxa – often only worms, midge larvae, leeches and water hog louse present in very high numbers – in worst case no life at all

The classification procedure depends upon the baseline pristine water data and this is sometimes difficult to obtain. A solution used in the UK is to take the physical information about the water and its catchment as inputs to a predictive model RIVPACS (river invertebrate prediction and classification system). This model then generates the number and types of taxa which would be expected in the type of water under assessment in the absence of pollution.

In 1976 the European Commission promulgated a directive concerned with the quality of surfacewaters intended for drinking water abstraction. This embodied the concept of a water quality objective for a specific use. As can be seen from

Table 2.4 (p. 23), the directive includes a wide range of physico-chemical and microbiological parameters which are given as guide levels and, for critical parameters, as mandatory levels which should not be exceeded.

Discharge consents

The direct discharge of wastewaters to watercourses is controlled by consent conditions related to the composition of the particular effluent. In relation to water pollution the Environment Agency maintains or improves water quality in receiving waters by controlling discharges to watercourses with legal documents known as consents. Charges are made by the Agency, on the basis of cost recovery, for setting consents and for monitoring of performance. Initial estimates of consent conditions are usually made by simple mass balance calculations but for significant discharges statistical and mathematical modelling techniques are increasingly employed. Each receiving water and discharge can have a wide range of flow and quality characteristics for which average values are of little significance. It is thus often necessary to establish frequency distributions for flow and important quality parameters in order to determine the effect of the discharge on the receiving water over a range of environmental conditions. Use of a Monte Carlo simulation technique can be appropriate for consideration of major discharges so that the probability of given water quality objectives being attained can be quantified.

Discharge consents, which may be reviewed after two years of operation, may be

- numeric – specifying concentrations of individual parameters as absolute values or as 95 percentile values when 'look-up' tables are used to judge compliance based on regular sampling programmes (95 percentile value is usually about twice the mean concentration for a series of samples)
- non-numeric – where environmental acceptability is defined by aesthetics rather than by concentrations
- descriptive – for small discharges where potential for environmental damage is slight and the type of treatment required is specified

Many current numeric discharge consents specify only two or three parameters, usually BOD, SS and ammonia but concerns arising out of several recent pollution incidents have indicated the value of an overall requirement that in addition to the numeric parameters includes wording such as

'. . . and shall not include concentrations of any other constituent which could be detrimental to other normal uses of the receiving water.'

This additional requirement would probably reduce the number of pollution incidents although it would place an added responsibility on dischargers to ensure that compliance was achieved.

An important area of water pollution control relates to prevention of pollution by industrial wastewater discharges. These may be made directly to a receiving water, in which case they are subject to discharge consents based on river quality objectives. Such consents normally specify absolute limits rather than percentiles. Industrial wastewaters are, however, frequently discharged to municipal sewers for treatment in admixture with domestic sewage. In this case constraints (trade effluent consents) are placed, by the water services company or sewerage authority, on the content of the discharge to prevent hazards arising in the sewers and to avoid interference with the treatment processes. Table 7.8 gives some examples of typical standards for the discharge of industrial wastewaters to municipal sewerage systems. Industrialists are charged for effluents on the basis of volume to be treated and the amount of treatment necessary to comply with effluent consent conditions at the sewage treatment plant. Pretreatment of industrial effluents may be required prior to discharge to the sewers or may be economically attractive to the industry.

Table 7.8 Typical standards for the discharge of industrial wastewaters to sewers

Parameter (mg/l except as noted)	Maximum concentration in industrial discharge
pH (units)	6–11
Sulphate	1500
Sulphide	10
Cyanide	10
Ammonia	100
Chromium	50
Grease/oil	500
Settleable solids	1000
Petroleum spirit	0
Temperature (°C)	43

Charging schemes for the reception and treatment of industrial effluents are usually based on the Mogden formula concept which is of the form

$$\text{Charge/m}^3 = R + V + \frac{O_i}{O_s} B + \frac{S_i}{S_s} S \qquad (7.16)$$

where R = reception and conveyance charge (p/m^3),
V = volumetric and primary treatment charge (p/m^3),
B = biological oxidation charge (p/m^3) of settled sewage,
O_i = COD (mg/l) of industrial effluent after settlement,
O_s = COD (mg/l) of settled sewage,

S = treatment and disposal costs of primary sludge (p/m^3),
S_i = total SS (mg/l) of industrial effluent and
S_s = total SS (mg/l) of incoming sewage.

Typical figures for the above factors (1996 prices) are given in Table 7.9.

This type of charging scheme encourages the industrial dischargers to take steps to reduce the volume and strength of the wastewater by careful process control and, possibly, modification of processes. The 'polluter must pay' policy sometimes advocated for dealing with industrial waste discharges may not be altogether satisfactory unless the charges are rationally based. In some situations an industrialist may prefer to pay the cost of causing pollution as an operating expense rather than having capital invested in a treatment plant. Such an approach would be likely to have generally detrimental effects on water quality.

Table 7.9 Average factors for industrial effluent charges (England and Wales 1996)

Factor in equation 7.16	1996 average value
R	15.3 p/m^3
V	13.8 p/m^3
B	19.1 p/m^3
S	12.5 p/m^3
O_s	542 mg/l
S_s	351 mg/l

Although in the UK the water quality objective approach is seen as the most logical way of achieving water pollution control and environmental protection, all EU states are subject to a directive on standards for wastewater treatment which specifies minimum levels of treatment for various circumstances (Table 7.10). These requirements will apply to all urban wastewater discharges in the

Table 7.10 Main requirements of the urban waste water treatment directive (91/271/EEC)

Type of receiving water	Effluent standard or removal requirements
Normal	BOD 25 mg/l MAC or 70–90% removal COD 125 mg/l or 75% removal
Sensitive areas (low dilution)	Total P 1–2 mg/l MAC or 80% removal Total N 10–15 mg/l MAC or 70–80% removal (Lower MACs for >100 000 population equivalents)
High natural dispersion (estuary/sea)	BOD 20% removal SS 50% removal

EU with implementation dates of up to 2005. For population equivalents <2000 for inland and estuary discharges and <10 000 for coastal discharges only 'appropriate treatment' is required. The directive also applies to organic industrial wastewaters which are discharged directly to natural waters. Adoption of the directive in the UK has produced examples of where such a fixed emission standard method of control appears to result in excessive regulation which cannot be justified on a cost–benefit analysis.

A number of other countries have adopted emission standards as primary controls on effluent discharges and Table 7.11 shows the situation in Japan where national effluent standards are related to health-related constituents and to environmentally important constituents. The national standards are taken as the minimum acceptable level and individual prefectures can set more stringent standards for their own particular requirements.

Table 7.11 Japanese effluent standards

Parameter (mg/l except where noted)	National standard	Typical prefecture standard
Health related parameters		
Cadmium	0.1	0.01
Cyanide	1.0	Not detectable
Organophosphorus compounds	1.0	Not detectable
Lead	1.0	0.1
Chromium (hexavalent)	0.5	0.05
Arsenic	0.5	0.05
Mercury	0.005	0.0005
Alkyl compounds	Not detectable	Not detectable
PCB	0.003	Not detectable
Environmentally related parameters		
BOD	160	25
SS	200	50
Phenol	5	0.5
Copper	3	1
Zinc	5	3
Iron (soluble)	10	5
Manganese (soluble)	10	5
Fluoride	15	10
Mineral oil	5	3
Fat	30	10
pH (units)	5.8–8.6	5.8–8.6
Coliforms (MPN/ml)	3000	3000

In the USA the effluent 'national pollutant discharge elimination system' (NPDES) standards are intended to protect and preserve beneficial uses of a receiving water, based on water quality criteria, technology-based limits, or both. Receiving water quality criteria are generally based on low-flow conditions which correspond to seven consecutive days with a ten year return period and take into account dilution and the characteristics of individual pollutants. National minima, known as secondary treatment equivalency, have been defined for municipal wastewater discharges and examples are set out in Table 7.12. Individual plants may be subject to additional requirements for advanced treatment where secondary treatment would not maintain the required receiving water quality.

Table 7.12 US minimum national standards for secondary wastewater treatment

Parameter	30 day average	7 day average
Five day BOD, most stringent of		
Effluent (mg/l)	30	45
Percentage removal	85	
Suspended solids, most stringent of		
Effluent (mg/l)	30	45
Percentage removal	85	
pH	Within range of 6.0 to 9.0 at all times	
Faecal coliforms (MPN/100 ml)	200	400

In the case of tidal waters, discharges may be regulated on the basis of the normal physical and chemical parameters used for inland discharges suitably adjusted to allow for the available dilution. Thus in situations with adequate dilution the discharge of screened or comminuted sewage may be acceptable. When the major concern is in relation to bathing or shellfish beds the bacteriological effects of sewage pollution are likely to be the most significant. Although the health implications of bacteriological contamination of tidal waters are difficult to quantify, various authorities have produced bathing water standards based on coliform counts ranging from 100/100 ml in California to 10 000/100 ml in the EU (Table 2.8, p. 26). Here again it would seem that local standards suited to the particular climatic and environmental conditions are more likely to be appropriate than universal standards.

Conventional wastewater treatment processes are not designed to remove harmful microorganisms so that a typical sewage effluent will contain large numbers of coliforms and other enteric bacteria. Where such effluents are discharged to bathing waters, either inland or coastal, it is becoming increasingly common to install a final disinfection stage to ensure compliance with bathing water quality standards. Ultraviolet disinfection is often seen as the only acceptable disinfection process for bathing waters because of concerns about possible disinfection by-products resulting from chemical disinfectants such as chlorine or ozone.

It must be appreciated that although the effective control of water pollution from point sources such as treatment plants and sewer overflows is reasonably well established these are not the only sources of water pollution. A considerable amount of pollution can occur via diffuse sources such as land drainage and surfacewater runoff from paved areas. As controls on point sources are tightened the pollution from diffuse sources can become more and more important. For example eutrophication can be caused by nutrients present in the effluents from sewage treatment plants so that nitrogen and phosphorus removal stages may be required. In some catchments this can prevent eutrophication but in agricultural catchments even complete removal of nutrients from sewage effluents may have little influence on eutrophication because of the nutrients present in runoff from farm land.

Drainage water from abandoned mines can have serious implications for water quality in some rivers where high iron concentrations and acidity levels can completely upset the natural balance. In industrialized areas runoff and seepage from contaminated land, sometimes as a consequence of activities which ceased many years ago, can carry toxic substances such as heavy metals and complex organics which can effectively render receiving waters unable to support life and prevent groundwater usage for water supply. In these situations control of pollution is often difficult and costly.

When efficient control of point sources is practised, non-point sources can contribute significant amounts of pollution and it is vital when preparing pollution-control policies to allow for this contribution since otherwise the benefits to the environment of the control policy may be overestimated.

References

Dobbins, W. E. (1964). BOD and oxygen relationships in streams. *Proc. Amer. Soc. Civ. Engnrs*, **90**, SA3, 53.

Gameson, A. L. H., Truesdale, G. A. and Downing, A. L. (1955). Re-aeration studies in a lakeland beck. *J. Instn Wat. Engnrs*, **9**, 571.

Klein, L. (1962). *River Pollution. 2. Causes and Effects*, p. 237. London: Butterworths.

Owens, S. M., Edwards, R. W. and Gibbs, J. W. (1964). Some reaeration studies in streams. *Int. J. Air Wat. Pollut.* **8**, 469.

Royal Commission on Sewage Disposal, Eighth Report, Cmd 6464. HMSO, London, 1912.

Further reading

Agg, A. R. and Zabel, T. F. (1989). EC Directive on the Control of Dangerous Substances (76/464/EEC): its impact on the UK water industry. *J. Instn Wat. Envir. Managt*, **3**, 436.

Balmforth, D. J. (1990). The pollution aspects of storm-sewage overflows. *J. Instn Wat. Envir. Managt*, **4**, 219.

Bicudo, J. B. and James, A. (1989). Measurement of reaeration in streams: Comparison of techniques. *J. Envir. Engng Am. Soc. Civ. Engnrs*, **115**, 992.

Biswas, A. K. (ed.) (1981). *Models for Water Quality Management*. New York: McGraw-Hill.

Burn, D. H. (1989). Water quality management through combined simulation–optimization approach. *J. Envir. Engng Am. Soc. Civ. Engnrs*, **115**, 1011.

Cole, J. A. (1974). *Groundwater Pollution in Europe*. New York: Water Information Center.

Department of the Environment (1986). *River Quality in England and Wales 1985*. London: HMSO.

Environment Agency (1997). *Groundwater Pollution*. Olton: EA.

Fiddes, D. and Clifforde, I. T. (1990). River basin management: developing the tools. *J. Instn Wat. Envir. Managt*, **4**, 90.

Foundation for Water Research (1994). *Urban Pollution Management (UPM) Manual*. Marlow: FWR.

Garnett, P. H. (1981). Thoughts on the need to control discharges to estuarial and coastal waters. *Wat. Pollut. Control*, **80**, 172.

Harryman, M. B. M. (1989). Water source protection and protection zones. *J. Instn Wat. Envir. Managt*, **3**, 548.

James, A. (ed.) (1993). *An Introduction to Water Quality Modelling*, 2nd edn. Chichester: Wiley.

Johnson, I. W. and Law, F. M. (1995). Computer models for quantifying the hydroecology of British rivers. *J. C. Intern Wat. Envir. Mangt*, **9**, 290.

Kay, D., Wyer, M., McDonald, A. and Woods, N. (1990). The application of water-quality standards to UK bathing waters. *J. Instn Wat. Envir. Managt*, **4**, 436.

Kinnersley, D. (1989). River basin management and privatization. *J. Instn Wat. Envir. Managt*, **3**, 219.

Kubo, T. (1991). Historical factors and recent developments in wastewater treatment in Japan. *J. Instn Wat. Envir. Managt*, **5**, 553.

Mance, G. (1993). Effluent and river quality: How the UK compares with other EC countries. *J. Instn Wat. Envir. Managt*, **7**, 592.

Martingdale, R. R. and Lane, G. (1989). Trade effluent control: prospects for the 1990s. *J. Instn Wat. Envir. Managt*, **3**, 387.

Mather, J. D. (1989). Groundwater pollution and the disposal of hazardous and radioactive wastes. *J. Instn Wat. Envir. Managt*, **3**, 31.

McIntosh, P. I. and Wilcox, J. (1979). Water pollution charging systems in the EEC. *Wat. Pollut. Control*, **78**, 183.

Miller, D. G. (1988). Environmental standards and their implications. *J. Instn Wat. Envir. Managt*, **2**, 60.

National Rivers Authority (1990). *Discharge Consent and Compliance Policy: A Blueprint for the Future*. London: NRA.

National Rivers Authority (1991). *Proposed Scheme of Charges in Respect of Discharges to Controlled Waters*. London: NRA.

National Rivers Authority (1991). *Proposals for Statutory Water Quality Objectives*. Bristol: NRA.

National Rivers Authority (1992). *Policy and Practice for the Protection of Groundwater*. Bristol: NRA.

National Rivers Authority (1992). *The Influence of Agriculture on the Quality of Natural Waters in England and Wales*. Bristol: NRA.

National Rivers Authority (1994). *The Quality of Rivers, Canals and Estuaries in England and Wales 1993*. Bristol: NRA.

National Rivers Authority (1994). *Water Quality Objectives*. Bristol: NRA.

National Rivers Authority (1994). *Contaminated Land and the Water Environment*. Bristol: NRA.

National Rivers Authority (1994). *Abandoned Mines and the Water Environment*. Bristol: NRA.

Newson, M., Marvin, S. and Slater, S. (1996). *Pooling Our Resources – A Campaigners' Guide to Catchment Management Planning*. London: CPRE.

Novotony, V. and Chesters, G. (1981). *Handbook of Nonpoint Pollution*. New York: Van Nostrand Reinhold.

Ross, S. L. (1994). Setting effluent consent standards. *J. Instn Wat. Envir. Managt*, **8**, 656.

Tebbutt, T. H. Y. (1979). A rational approach to water quality control. *Wat. Supply Managt*, **3**, 41.

Tebbutt, T. H. Y. (1990). *BASIC Water and Wastewater Treatment*, Chapter 4. London: Butterworths.

Velz, C. J. (1984). *Applied Stream Sanitation*, 2nd edn. New York: Wiley-Interscience.

Problems

1. A quiescent body of water has a depth of 300 mm and its DO concentration is 3 mg/l. Determine the DO concentration at the bottom after a period of twelve days if the surface is exposed to the atmosphere at a temperature of 20°C. At 20°C k_d is 1.86×10^3 mm^2/s. (4 mg/l)

2. A town of 20 000 people is to discharge treated domestic sewage to a stream with a minimum flow of 0.127 m^3/s and BOD 2 mg/l. The sewage dry weather flow is 135 l/person day and the per capita BOD contribution is 0.068 kg/day. If the BOD in the stream below the discharge is not to exceed 4 mg/l, determine the maximum permissible effluent BOD and the percentage purification required in the treatment plant. (12 mg/l, 97.5 per cent)

3. A stream with BOD 2 mg/l and saturated with DO has a normal flow of 2.26 m^3/s and receives a sewage effluent, also saturated with DO, of 0.755 m^3/s with BOD 30 mg/l. Determine the DO deficits over the next five days and hence plot the sag curve. Calculate the critical DO deficit and the time at which it occurs. Assume temperature is 20°C throughout. Saturation DO at 20°C is 9.17 mg/l, k_1 for effluent/water mixture is 0.17/day, k_2 for stream is 0.40/day. (2.38 mg/l, 1.61 days)

4. A stream with a flow of 0.75 m^3/s and BOD 3.3 mg/l is saturated with DO (9.17 mg/l at 20°C). It receives an effluent discharge of 0.25 m^3/s, BOD 20 mg/l and DO 5.0 mg/l. Determine the DO deficit at a point 35 km downstream if the average velocity of flow is 0.2 m/s. Assume temperature is 20°C throughout, k_1 for effluent/water mixture 0.10/day, k_2 for stream 0.40/day. (1.87 mg/l)

5. A stream with flow 4 m^3/s, BOD 1 mg/l and saturated with oxygen receives at A a sewage effluent discharge of 2 m^3/s with BOD 20 mg/l and DO 4 mg/l. At point B, 20 km downstream of A, a tributary with flow 2 m^3/s, BOD 1 mg/l and DO 8 mg/l joins the main stream. A further distance of 20 km downstream at C the stream receives another effluent of 2 m^3/s with BOD 15 mg/l and DO 6 mg/l. Determine the DO deficit to point D, 20 km downstream of C assuming constant temperature of 20°C for which the saturation DO is 9.1 mg/l. For all reaches of the stream, $k_1 = 0.1$/day and $k_2 = 0.35$/day, velocity of flow = 0.3 m/s. (2.2 mg/l)

8

Water demands and wastewater flows

The basic human physiological requirement for water is about 2.5 l/day, although when working hard in hot climates an intake of up to 20 l/day may be necessary to prevent dehydration brought about by losses through perspiration. As a person's standard of living increases, so does the consumption of water in the home, although this additional water is not normally ingested. Most of the water taken into a house is returned as wastewater and many industries discharge volumes of wastewater which are similar to their water consumption. Agricultural uses of water, particularly in developing countries, exert the largest demands on water resources.

8.1 Domestic water demand

In addition to the water required for survival, there are many domestic uses of water which are highly desirable, e.g. for personal hygiene, in cooking, washing of utensils and clothing, etc. The amount of water used domestically is governed by the life style of the community and the availability of water, judged by both quantity and cost. In very primitive cultures water demands of around the minimum 2.5 l/day have been found, but as life styles develop, a water demand of about 10 l/person day is normal in the absence of a piped supply. The availability of a central stand-pipe in a village will probably give rise to a consumption of 25–30 l/person day and the provision of a single house tap will increase the demand to about 50 l/person day. In developed countries and in high-value urban housing in developing countries the provision of multiple taps, flush toilets, washing machines and dishwashers, large gardens and swimming pools can greatly increase domestic water demands to levels of several hundred litres per person day. Table 8.1 gives information about typical domestic water usage in the UK which indicates that a reasonable value for domestic consumption in a temperate developed country would be 140–150 l/person day. The flush toilet accounts for a considerable proportion of domestic water consumption and although it has its hygienic attractions it does use a great deal of water to carry human wastes through necessarily large sewers to the treatment works where the polluting matter is concentrated again at considerable expense.

Table 8.1 Typical domestic water usage in the UK

Use	Consumption (l/person) day
Toilet flushing	50
Baths and showers	35
Cooking and dishwashing	25
Clothes washing	20
Garden watering	7
Drinking	3
Car washing	2
Total	**142**

The low cost of water and its ready availability in some parts of the world make water carriage of wastes attractive, but it is not a technique which scores highly in relation to sustainability of resources. In the UK, new property is fitted with low-flush cisterns which use a 6 l flush. Earlier attempts to use dual-flush cisterns which provided a 9 l flush after defaecation but only a 4 l flush after urination were largely unsuccessful due to incorrect usage and frequent malfunctions. More recent toilets are designed to operate effectively with a 6 l flush. In some parts of the world much larger flushes are often employed so that in the USA and urban areas in tropical countries, for example, large flush volumes and the demands of air conditioners and garden watering can produce domestic water consumptions in excess of 500 l/person day.

8.2 Industrial water demand

Most industrial operations consume some water in addition to that required by the workers and in manufacturing areas the industrial water demand may equal or exceed the domestic demand. Table 8.2 shows some typical industrial water consumptions which are, however, greatly influenced by such factors as the age of the plant, the cost of water and the potential and incentive for in-plant recycling. Many industrial uses of water do not require a supply of potable quality and there is increasing use of lower grade supplies such as raw river water and sewage effluents. Industrial water consumption is closely related to industrial productivity so that it may change quite rapidly as influenced by economic circumstances. Although water is still a relatively cheap raw material the increasing cost of water supplies and the treatment of wastewaters has encouraged many industrial users to reduce their water intake. Water consumption and energy usage are now being increasingly recognized by industry as expenditures which are controllable. Clean technology concepts are bringing

Table 8.2   Examples of industrial water usage

Product or service	Consumption	Units
Coal	250	l/tonne
Bread	1.3	l/kg
Meat products	16	l/kg
Milk bottling	3	l/l
Brewing	5	l/l
Soft drinks	7	l/l
Chemicals	5	l/kg
Steel rolling	1 900	l/tonne
Iron casting	4 000	l/tonne
Aluminium casting	8 500	l/tonne
Automobiles	5 000	l/vehicle
Electroplating	15 300	l/tonne
Carpets	34	l/m^2
Textile dyeing	80	l/kg
Concrete	390	l/m^3
Paper	54 000	l/tonne
Dairy farming	150	l/cow day
Pig farming	15	l/pig day
Poultry farming	0.3	l/bird day
Schools	75	l/person day
Hospitals	175	l/person day
Hotels	760	l/employee day
Shops	135	l/employee day
Offices	60	l/employee day

reductions in water consumption and in wastewater production with benefits to both industry and the environment.

Within Europe 53 per cent of water abstracted is used for industrial purposes, 26 per cent is utilized in agriculture and 19 per cent is consumed domestically. With increasing demands on food production, particularly in developing countries, agricultural use of water for irrigation purposes is by far the largest use of water on a worldwide basis at around 65 per cent of total demand. However, many irrigation systems are highly inefficient in their use of water with relatively small amounts of the water actually reaching the plants. Much attention is now being directed to more effective irrigation practices such as drip-feed where small amounts of water can be delivered directly to individual plants.

8.3 Demand management

In many countries all water consumers are said to be metered, although in apartment blocks there is often only a single meter and the bill is divided amongst individual apartments. The UK does not have a general policy of universal metering for domestic consumers, but all significant industrial consumers are supplied by meter. However, increasing pressures on water resources together with a view that water charges should where possible be based on actual usage have encouraged the gradual introduction of domestic water meters. New domestic properties in the UK are fitted with meters and existing customers are encouraged to install meters at little or no cost. Nevertheless, less than 10 per cent of domestic consumers are supplied by meter and the majority pay for water on a fixed charge basis which is essentially related to the size and value of the property. Unmetered customers thus have no incentive to economize in the use of water. It has long been argued that the cost of installing and reading domestic water meters would far outweigh the value of any possible savings in consumption which might accrue. Water is a cheap product which in the UK is currently delivered for about £0.6/tonne and the initial cost of installing a meter is in excess of £200 depending upon its location. Evidence from countries where domestic metering is practised is sometimes conflicting and it is by no means certain that the introduction of universal domestic metering would produce a sustained significant reduction in consumption. Large-scale trials were under-taken at several locations in the UK to determine the costs and feasibility of domestic metering with the additional aim of assessing the value of meters in demand management. The results of the trials indicated that the use of meters tended to reduce domestic water consumption by up to 10 per cent although whether this reduction would be maintained in the longer term is not certain. As with other utilities like electricity and gas the water distribution system must be sized for the maximum demand which may only occur for a short time on a few days in the year. Sophisticated meters capable of real-time recording would permit differential pricing to discourage usage at peak times of demand but it is doubtful whether this level of complexity would be viable in the UK.

A much simpler method of discouraging excessive usage of water, which requires only a totalizing meter, is to charge for what is accepted as a reasonable domestic consumption at one rate with a significant surcharge for demands above the norm. This would tend to discourage excessive garden watering in times of possible water shortage without penalizing normal consumption. It is important to remember that an ample supply of safe water is a primary element in the maintenance of public health and thus poorer members of the community should not be forced to restrict essential water usage by economic measures. Demand management policies for water supply systems are aimed at affecting demand patterns so that they can be met by the resources available. The installation of meters and the adoption of seasonal tariffs are only part of the demand management scenario. It is worth noting that in the UK metering trials the installation of meters did appear to have a significant effect on peak demand rates, often caused by garden watering, with reductions of between 15 and 30 per cent in peak rates being observed generally.

It must be appreciated that in most water distribution systems some 25 per cent of the water is unaccounted for and is lost due to leakage, waste and unauthorized connections. The accurate evaluation of losses and leakage is made difficult in the UK because of the absence of domestic water meters. Thus the total unaccounted for water can only be obtained by subtracting known metered demands and estimated domestic consumption from the measured inputs to a distribution system. There is clearly a margin of error in such a calculation although by careful night-time metering of distribution zones it is possible to arrive at what are probably reasonably accurate measures of leakage and waste. However, even in countries where distribution systems are claimed to be fully metered there are often leakage levels in excess of 25 per cent, sometimes as high as 50 per cent in developing countries. The amount of leakage in a distribution system is greatly affected by its age and complexity so that in general a new system in a modern development might easily achieve an unaccounted for water level of less than 10 per cent. On the other hand a 100 year old system in a run-down urban area might well produce leakage rates in excess of 35 per cent. A common measure of leakage, determined by night-time district metering surveys, is a figure expressed as litres per property per hour with an average UK value of about 12 and a typical range of 5–20 l/property h.

The control of distribution pressures can play a major part in reducing leakage because, without accurate pressure control, the low flows at night time allow high pressures in the distribution system which increase leakage flows. Pressure management aims to prevent mains pressures from exceeding the minimum service level, usually 40–50 metres head in the UK, at any time. Effective enforcement of pressure management may well be able to reduce leakage by up to 30 per cent at low cost and at no significant disadvantage to the consumer. Many operators of water distribution networks would now aim to reduce leakage to around 10 per cent, below which level the cost of further work is likely to be much more than the value of the saved water.

Pressure reduction and leakage control strategies are thus important parts of a demand management strategy. Occasional bans on hose pipes and garden sprinklers may cause minor annoyance to consumers but they serve to reduce demands on the system which might otherwise require considerable investment resulting in increased water charges which would annoy consumers even more! Demand management in relation to both average and peak demands can delay the need for investment in additional resources and encourage sustainable use of existing resources. Additional measures to reduce demands on scarce water resources could include

- installation of low-volume taps and plumbing fittings
- use of 'grey' water for toilet flushing
- on-site collection of rainwater for toilet flushing and garden watering
- development of domestic appliances with low water consumption
- recycling of wastewater effluents, possibly in dual-supply networks

Although all of the above proposals are capable of reducing domestic water consumption to some extent they do have inherent disadvantages due to installation and operating costs and, for some, the potential for water pollution from cross connections between potable and secondary supply systems.

Treatment plants and their associated collection and distribution systems are costly items of capital expenditure which are thus designed to have long useful lives. Many treatment facilities will have design lives of at least 30 years and the underground pipes and sewers often remain in use for over 100 years. Because of these characteristics and the need to develop and utilize water resources in an efficient manner it is necessary to be able to predict future water demands as accurately as possible. Domestic water demand is the product of the individual per capita demand and the population served. The per capita demand tends to approach a ceiling value which is related to the standard of living of a community and its environmental conditions. In the UK and most Northern European countries it appears that in-house domestic use of water has altered little over the past 10 years but garden-watering demands have increased. Thus the total domestic per capita consumption probably increases at around 1 per cent per year.

It is, however, worth noting that the designs of such household appliances as washing machines and dishwashers are now placing more emphasis on energy and water saving features. It is thus possible that the future may show actual decreases in domestic water consumption in areas with a high standard of living where the population tends to be more environmentally conscious than elsewhere. Industrial demands for water in countries like the UK may be similar in size to the domestic demand but they are very much influenced by economic factors. The economic prosperity of an area is influenced by industrial productivity and this is reflected in industrial water consumption. Changes in the nature of industry in an area can result in sudden alterations in water demands which are difficult to predict and which can have major financial consequences for the water undertaking.

8.4 Population growth

The classical biological growth curve was discussed in Chapter 6 and although this may be considered as a basis for predicting the growth of human populations many factors make this model of limited use for such purposes. In most of the developed world the growth in population is now more or less negligible but the situation is very different in less developed countries where populations can double in a period of a few years, as shown in Figure 8.1. Improved living standards and better medical care in developed countries have increased life expectancy by some 50 per cent during the past 100 years. Variations in national or local economies can affect the birth rate as do wars and civil unrest. Changes in industry can completely alter the nature of growth in an area and the decline of an industry can cause a fall in population in the surrounding area.

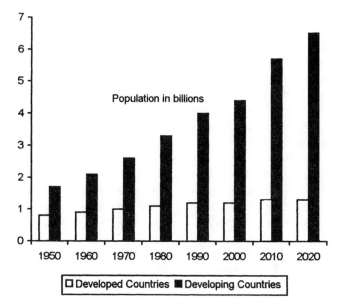

Figure 8.1 World population growth.

In most parts of the world a census is taken at ten-year intervals and it is logical to use this information as an aid in population prediction although its accuracy in developing countries may be dubious. Various techniques can be employed to predict future populations from existing records. If the growth is believed to be linear the following expression can be used

$$Y_m = \frac{t_m - t_2}{t_2 - t_1} (Y_2 - Y_1) + Y_2 \tag{8.1}$$

where Y_m = population in future year t_m. Y_1 and Y_2 are populations in years t_1 and t_2.

If geometric growth is believed to be appropriate, equation 8.1 can be modified to

$$\log Y_m = \frac{t_m - t_2}{t_2 - t_1} (\log Y_2 - \log Y_1) + \log Y_2 \tag{8.2}$$

Some authorities prefer to adopt a procedure of fitting complex polynomial expressions to census data. The use of mathematical relationships may, however, give a spurious accuracy to predictions and in many cases the best method is to plot census data and extrapolate the line for the required design period using all the information available regarding future developments in the area.

8.5 Wastewater flow

In temperate climates most of the water used in the home and in industry finds its way back into the sewers. Losses in garden watering and other consumptive uses are likely to be compensated for by groundwater infiltration which occurs in most sewerage systems. Infiltration of groundwater into sewers can significantly increase flows, particularly in older sewerage systems in poor condition. A typical design allowance for infiltration would be at least 10 per cent of the per capita water consumption in the area although the real figure is highly site specific and can range between 15 and 50 per cent of the dry weather flow. Thus the flow of wastewater in dry weather is likely to be of the same order as the flow of water supplied to an area. In hot climates a significant proportion of the water supplied will be used in garden watering or otherwise lost by evaporation so that only 70–80 per cent may enter the sewers.

The volume and nature of wastewater depends upon the type and age of the sewerage system. In old systems, damaged sewers and cracked joints may allow loss of sewage into the surrounding ground and conversely infiltration of groundwater may increase the flow of sewage. Older communities usually have combined sewers which convey both the foul sewage, from baths, sinks, WCs, etc., and the surfacewater runoff due to rainfall on paved areas and roofs. Even in moderate rainfalls, the surfacewater runoff becomes much larger than the flow in dry weather from a built-up area and sewers would need to be uneconomically large to contain the flow. It is therefore customary to install combined sewer overflows (CSOs) which divert flows in excess of 6, 9 or sometimes 12 times the dry weather flow to a nearby water course. This inevitably causes significant pollution, although hopefully the flow in the water course receiving the storm discharge will be high because of the rainfall. A major problem with CSO discharges is the release of paper, plastics and sanitary items which are aesthetically unattractive on the banks of the receiving water. With combined sewers there are hydraulic design problems related to the need to maintain a minimum self-cleansing velocity at low flows whilst preventing excessive velocities when the sewer is running full. It should be noted that the achievement of self-cleansing velocities in sewers is particularly important in tropical areas since in those conditions organic deposits will rapidly become anaerobic and the resultant production of hydrogen sulphide can cause serious damage to the sewer. Because of the disadvantages of combined sewers, most new developments are sewered on a separate system with one relatively small foul sewer, all of whose contents are treated, and with storm water sewers which carry only relatively clean runoff and which can safely be discharged to local water courses. The cost of a separate system will inevitably be somewhat higher than that of a combined system although in many cases it may be possible to lay both pipes in the same excavation for at least part of their length.

In the UK dry weather flow (d.w.f.) is usually defined as

$$\text{d.w.f.} = PG + I + E \tag{8.3}$$

where P = population served,

G = average domestic wastewater contribution (m^3/person day),

I = infiltration (m^3/day) and

E = industrial wastewater discharged over 24 hours (m^3).

Combined sewer overflow (CSO) settings were traditionally based on the Ministry of Housing and Local Government formula

$$Q = \text{d.w.f.} + 1.36P + 2E \tag{8.4}$$

where Q = flow to full treatment and the other terms have the same meanings as in equation 8.3.

The adoption of the UPM procedures mean that for all but small schemes a more rational approach to the setting of CSO discharge levels can be used in preference to the pragmatic approach which is the basis of equation 8.4.

The rate of surface runoff depends upon the intensity of rainfall and the impermeability of the area drained which varies according to its nature (Table 8.3). Rainfall intensity varies with the return period of the storm and the duration for which it lasts and various empirical relationships for rainfall intensity are available for different geographical locations.

The Bilham formula can be used in the UK to provide a first estimate of rainfall intensity

$$R = \left(\frac{14.2F^{0.28}}{t^{0.72}} - \frac{2.54}{t} \right) \tag{8.5}$$

where R = rainfall intensity (mm/h), F = return period of storm (years) and t = storm duration (hours).

Table 8.3 Typical impermeability factors

Surface	Impermeability factor
Watertight roof	0.70–0.95
Asphalt pavement	0.85–0.90
Concrete flagstones	0.50–0.85
Macadam road	0.25–0.60
Gravel drive	0.15–0.30
Undeveloped land	
flat	0.10–0.20
sloping	0.20–0.40
steep rocky slope	0.60–0.80

The maximum runoff from an area is obtained when the duration of the storm is equivalent to the time of concentration of the area. The time of concentration has two components, the time of entry, i.e. the time for rain to flow along gutters and drainpipes into the sewer, and the time of flow along the sewer.

The runoff from a drainage area is given by

$$Q \text{ (m}^3\text{/s)} = 0.278 \, A_p R \tag{8.6}$$

where A_p = impermeable area (km^2) and R = rainfall intensity (mm/h).

More detailed estimates of the rainfall intensity and the storm profile are used for major urban surfacewater sewerage schemes where designs are usually undertaken with the aid of some form of hydrological model of which the Wallingford procedure is the most popular in the UK. This procedure is based on the results of the NERC Flood Studies Report which enables urban runoff flows to be predicted on the basis of catchment characteristics, rainfall event profiles and return period. The diameter and gradient of pipes in the drainage system can then be determined using an optimizing routine if required.

8.6 Variations in flow

Although the average water demand and sewage dry weather flow (d.w.f.) may be determined for a community as say 150 l/person day there will be considerable variations over a 24 h period. The magnitude of these variations depends upon the size of the population concerned. In the case of water consumption, the ratio of peak hourly rate to annual average rate can vary from about 3 for a population of a few hundred to about 2.2 for a community of 50 000 and about 1.9 for a population of half a million. In hot weather the use of garden sprinklers can produce very high peak hourly demand rates, sometimes of up to six times the average rate. Somewhat similar variations are to be expected with the d.w.f. in sewers although the longer the sewer system the greater the smoothing effect on flow variations. The effects of surface runoff will of course greatly amplify the peak flows in combined systems. Even in separate systems, flows can be affected by rainfall since illegal gulley connections and small areas of impermeable surfaces are often served by the foul sewer.

Water-treatment plants can often operate at a constant output with the distribution system and service reservoirs serving to balance the fluctuating demand. However, differential electricity tariffs may encourage higher rates of treatment during the night. In the case of wastewater-treatment plants the flow of sewage arriving at the works will be balanced to some extent within the sewerage system, but the plant must be designed to operate under fluctuating flows with a normal maximum capacity for full treatment of

$$FTF = 3PG + I + 3E \tag{8.7}$$

where FTF = full treatment flow and other symbols are as for equation 8.3.

Further reading

Bartlett, R. E. (1979). *Public Health Engineering: Sewerage.* 2nd edn. Barking: Applied Science Publishers.

Bartlett, R. E. (ed.) (1979). *Developments in Sewerage, Vol.* 1. Barking: Applied Science Publishers.

Central Water Planning Unit (1976). *Analysis of Trends in Public Water Supply.* Reading: CWPU.

Department of the Environment (1992). *Using Water Wisely.* London: DOE.

Department of the Environment (1995). *Water Conservation – Government Action.* London: DOE.

Dovey, W. J. and Rogers, D. V. (1993). The effect of leakage control and domestic metering on water consumption in the Isle of Wight. *J. Instn Wat. Envir. Managt,* **7**, 156.

Edwards, A. M. I. and Johnston, N. (1996). Water and waste minimization in the Aire and Calder project. *J. C. Instn Wat. Envir. Managt,* **10**, 227.

Gadbury, D. and Hall, M. J. (1989). Metering trials for water supply. *J. Instn Wat. Envir. Managt,* **3**, 182.

Gilbert, J. B., Bishop, W. J. and Weber, J. A. (1990). Reducing water demand during drought years. *J. Am. Wat. Wks Assn,* **82**(5), 34.

Males, D. B. and Turton, P. S. (1979). *Design Flow Criteria in Sewers and Water Mains.* Reading: CWPU.

Mills, R. E. (1990). Leakage control in a universally metered distribution system: Pinetown Water's experience. *J. Instn Wat. Envir. Managt,* **4**, 235.

National Rivers Authority (1994). *Water – Nature's Precious Resource.* London: HMSO.

National Rivers Authority (1995). *Saving Water.* Bristol: NRA.

Office of Water Services (1996). *Report on Recent Patterns for Water Demand in England and Wales.* Birmingham: OFWAT.

Shaw, Elizabeth M. (1990). *Hydrology in Practice,* 2nd edn. London: Chapman and Hall.

Shore, D. G. (1988). Economic optimization of distribution leakage control. *J. Instn Wat. Envir. Managt,* **2**, 545.

Smith, A. L. and Rogers, D. V. (1990). Isle of Wight water metering trial. *J. Instn Wat. Envir. Managt,* **4**, 403.

Smith, E. D. (1997). The balance between public water supply and environmental needs. *J. C. Instn Wat. Envir. Managt,* **11**, 8.

Sterling, M. J. H. and Antcliffe, D. J. (1974). A technique for prediction of water demand from past consumption data. *J. Instn Wat. Engnrs,* **28**, 413.

Thackray, J. E. (1992). Paying for water: Policy options and their practical implications. *J. Instn Wat. Envir. Managt,* **6**, 505.

Thackray, J. E. and Archibald, G. G. (1981). The Severn–Trent studies of industrial water use. *Proc. Instn Civ. Engnrs,* **70**(1), 403.

Thackray, J. E., Cocker, V. and Archibald, G. G. (1978). The Malvern and Mansfield studies of domestic water usage. *Proc. Instn Civ. Engnrs,* **64**(1), 37.

Water Services Association (1996). *Waterfacts '96.* London: WSA.

White, J. B. (1986). *Wastewater Engineering,* 3rd edn. London: Edward Arnold.

Problems

1. Using the census data below for two communities predict the populations in 2001 using arithmetic, geometric and graphical techniques.

Year	Developed-country town	Developing-country city
1921	64 126	257 295
1931	67 697	400 075
1941	69 850	632 136
1951	74 024	789 400
1961	75 321	1 227 996
1971	75 102	2 022 577
1981	73 986	2 876 309
1991	74 390	3 723 467

(There are no correct answers for this problem and those obtained will depend on how much of the existing data are utilized.)

2. Determine the rainfall intensity as given by the Bilham formula for storms of 15 min duration with return periods of 1, 5 and 10 years. (28.4, 50.4 and 63.5 mm/h)

3. Using the rainfall intensities determined above calculate the corresponding runoff values for an area of 3 km^2 which has an average impermeability factor of 0.45. (10.7, 18.9 and 23.8 m^3/s)

9

Introduction to treatment processes

It will be apparent from previous chapters that waters and wastewaters often have highly complex compositions and that modifications to the composition are usually necessary to suit a particular use or to prevent environmental degradation. It follows that a variety of treatment processes will be necessary to deal with the range of contaminants likely to be encountered.

Contaminants may be present as

- floating or large suspended solids: in water – leaves, branches, etc.; in wastewater – paper, rags, grit, etc.
- small suspended and colloidal solids: in water – clay and silt particles, microorganisms; in wastewater – large organic molecules, soil particles, microorganisms
- dissolved solids: in water – alkalinity, hardness, organic acids; in wastewater – organic compounds, inorganic salts
- dissolved gases: in water – carbon dioxide, hydrogen sulphide; in wastewater – hydrogen sulphide
- immiscible liquids, e.g. oils and greases.

The actual particle size at which the nature of the material changes from one group to another depends upon such physical characteristics as specific gravity of the material and the division between groups is, in any event, indistinct. In certain cases it may be necessary to add substances to the water or wastewater to improve its characteristics, e.g. chlorine for disinfection of water, oxygen for the biological stabilization of organic matter.

9.1 Methods of treatment

There are three main classes of treatment process (summarized below), typical operational ranges of which are shown in Figure 9.1.

1. Physical processes, which depend essentially on physical properties of the impurity, e.g. particle size, specific gravity, viscosity, etc. Typical examples of this type of process are screening, sedimentation, filtration, gas transfer.

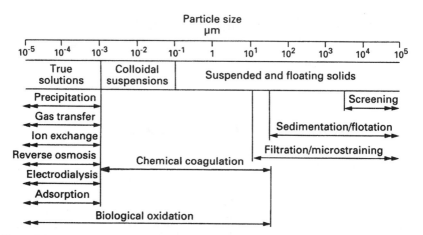

Figure 9.1 Applications of the main treatment processes.

2. Chemical processes, which depend on the chemical properties of an impurity or which utilize the chemical properties of added reagents. Examples of chemical processes are coagulation, precipitation, ion exchange.
3. Biological processes, which utilize biochemical reactions to remove soluble or colloidal impurities, usually organics. Aerobic biological processes include biological filtration and activated sludge. Anaerobic oxidation processes are used for the stabilization of organic sludges and high strength organic wastes.

In some situations, a single treatment process may provide the desired change in composition but in most cases it is necessary to utilize several processes in combination. For example, sedimentation of a river water will remove some, but by no means all, of the suspended matter. The addition of a chemical coagulant followed by gentle stirring (flocculation) will cause the agglomeration of colloidal particles which can then largely be removed by sedimentation. Most remaining non-settleable solids can be removed by filtration through a bed of sand. The addition of a disinfectant serves to kill any harmful microorganisms which have survived the preceding stages of treatment.

The probable combinations of treatment processes required to produce potable water from various sources are given in Table 9.1 and Table 9.2 shows typical domestic sewage treatment systems for various effluent qualities. Flow sheets and typical design criteria for conventional water-treatment and domestic sewage-treatment plants appear in Figures 9.2 and 9.3 respectively.

9.2 Optimized design

As outlined above, treatment plants usually consist of a number of unit processes or operations in combination. Most plants are designed using fairly standard

Table 9.1 Probable treatment for various raw waters

Source	Probable treatment	Possible additions
Upland catchment	Screening or microstraining, disinfection	Sand filtration Stabilization Colour removal
Lowland river	1. Screening or microstraining, coagulation, rapid filtration, disinfection 2. Screening or microstraining, rapid filtration, slow filtration, disinfection	Storage Softening Stabilization Adsorption Desalination Nitrate removal
Deep groundwater	Disinfection	Softening Stabilization Iron removal Desalination Nitrate removal

criteria, of the type shown in Figures 9.2 and 9.3, which have been developed over the years and which will usually produce satisfactory levels of performance. However, it is important to appreciate that such an approach usually tends to lead to a somewhat conservative design. A more rational approach is based on the concept of the treatment units forming a system in which each unit is designed to perform a particular function and the overall system is optimized economically. Increasing capital and operational costs mean that investments in treatment plants must be scrutinized carefully to ensure that the best value for money is

Table 9.2 Probable domestic sewage treatment for various receiving waters

Receiving water	Typical effluent standard (mg/l)		Probable treatment
	BOD	SS	
Open sea	–	–	Screening or maceration, or as for tidal estuary below, depending upon legislation
Tidal estuary	150	150	Screening followed by primary sedimentation with sludge disposal on land or by incineration
Lowland river	20	30	Screening followed by primary sedimentation, aerobic biological oxidation, secondary sedimentation, sludge stabilization and sludge disposal on land or by incineration
High-quality river	10	10	As for lowland river with addition of tertiary treatment by sand filtration, grass plots, reed beds or lagoons

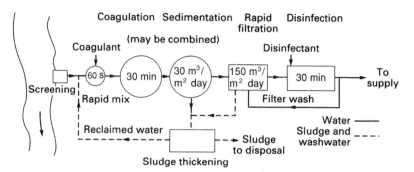

Figure 9.2 Conventional water-treatment plant with typical design criteria indicated.

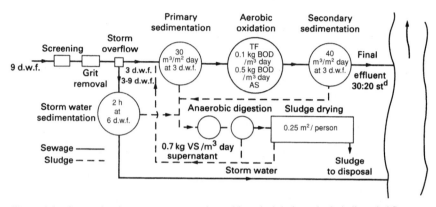

Figure 9.3 Conventional sewage-treatment plant with typical design criteria indicated. AS = activated sludge; TF = tracking filter; VS = volatile solids.

obtained. Whilst it is true that the effective operation of water and wastewater treatment processes can have a highly beneficial effect on public health, such activities should not be shielded from rational analysis.

The use of systems analysis concepts to develop mathematical optimizing models of treatment plants can provide a useful aid to the designer provided that reliable performance and cost data are available. A primary requirement for optimization is the availability of performance relationships for each unit process, linking input and output qualities with a characteristic loading parameter. Performance relationships may be established on the basis of knowledge of the theoretical behaviour of the particular process or on the basis of an empirical model for the process. In either case it is necessary to prove that the model developed does provide a satisfactory representation of the process for which it has been produced.

The costs of treatment, both capital and operating, are important factors in any design but the establishment of reliable cost functions is not easy because of site-specific conditions. Wide variations in capital costs are found at different plants since ground conditions or site configuration can have a marked effect on costs and make comparison with data from other plants somewhat difficult. There are, however, clear economies of scale with most treatment facilities so that the per capita cost of water or wastewater treatment for a village will be several times greater than for a city.

By combining performance relationships and cost functions it is possible to produce a mathematical model of a complete treatment plant which can be used by a designer to evaluate a number of treatment options and thus arrive at the optimum design. A number of commercial treatment plant models are now available of which STOAT produced by the Water Research Centre (WRC) is a good example of a dynamic performance simulation model.

9.3 Control and operation

Water and wastewater treatment plants involve a number of inter-related processes and operations which can be carried out manually but which are increasingly being undertaken by automatic control systems. The growth of information technology (IT) has encouraged the adoption of SCADA (supervisory control and data acquisition) systems which can enable complex plants and even groups of plants to be operated from a central location by small numbers of supervisory staff. An extension to this concept is to incorporate intelligent knowledge-based systems into the control framework so that in most circumstances human intervention is not required. It is important to appreciate that such IT systems can be costly to purchase and they require highly skilled staff for their installation, calibration and maintenance.

There are three basic options for a process plant, described below.

1. Completely manual operation and control in which all decisions and adjustments are undertaken by human operatives. Individual valves and other controls are operated manually at their locations as required by flow measurements and/or water quality measurements.
2. Manual operation with automated control in which decisions are made by human operatives who manually initiate the operation of valves and other controls from a central location.
3. Fully automated operation and control in which all normal decisions are made and acted upon by local or central programmed logic controllers integrated into an intelligent system.

In broad terms, these options have increasing capital costs and decreasing manpower needs although the fully automated system will require a small number of highly qualified staff for its maintenance and to take manual control in the event of system malfunction or failure.

A completely manual system would involve adjusting valves, starting or stopping pumps and other mechanical equipment, altering chemical dosing, desludging tanks and backwashing filters as indicated by oral or written information on flow or quality. Such a system would use local operation and thus require regular visits to various parts of the plant. This has the benefit of providing visual inspection of treatment units but may be unpopular with workers at night or in bad weather when safety hazards may exist.

An automated manual system reduces the need for staff by collecting information on flow and quality, often from remote continuous sensors, and presenting relevant information to a centrally located controller who can then alter or adjust individual parts of the plant by remote control without visiting their location. This provides a better integrated operation of a plant and reduces the need for regular visits to remote parts of the plant. Alarm systems are essential to indicate if control actions initiated by the controller have not been actioned at the remote location.

In a fully automated system, information on flow and appropriate quality parameters together with status reports on individual components is supplied to a central computer. This central computer, often assisted by local programmed logic controllers, uses models of the plant components to enable decisions to be reached and actioned. Such systems need to be carefully set up and calibrated with fail-safe operating controls and alarms to indicate the need for human override. The most advanced automated systems include a learning mode which permits them to add to their intelligence and reduce the need in the future for human intervention. Early automated systems were somewhat limited because decisions were reached on a 'yes/no' basis. Newer systems are able to use 'fuzzy logic', as does the human brain, to arrive at decisions which are 'probably' or x per cent correct, given the information received.

In essence, process control implies the use of information to maintain flow or quality parameters within specified limits. Thus a person observing a steady fall in a flow meter indication can adjust a valve to return the flow to its required value. Equally a microchip can achieve the same objective on the basis of instructions with which it has been supplied. There are two basic types of control system.

1. Feedforward control in which measurements in the feed to a process are used to adjust the process conditions by means of a model which uses the measured value(s) to predict the process state. The model then determines the change in regulator position (valve opening, chemical dose, etc.) to achieve the required value of the control variable.
2. Feedback control uses measurements in the output to make appropriate adjustments to the process state. The controller determines an error value by comparing the measured and required values of the control variable. The correction value produced is a function of the size of the error and the effect of the control signal is transmitted back to the controller as the process variable returns to the required value.

With feedback control three options are available

1. Proportional, where the correction signal is proportional to the error
2. Integral, where the correction signal is proportional to a time-integrated value of the error
3. Derivative, where the correction signal is proportional to the rate of change of the error.

Simple proportional control may result in an equilibrium value which is offset from the required value and a sensitive system can oscillate or 'hunt' about the required value with wide variations. Proportional plus integral control ensures that adjustments continue as long as an error exists so that even small errors can be reduced and offset eliminated. Proportional plus derivative control speeds up the control action by anticipating future errors and taking pre-emptive action. This reduces oscillations but with constant errors an offset equilibrium may occur. Proportional plus integral plus derivative control (three-term control) uses integral action to eliminate offset and derivative action to speed up response to deviations and decrease oscillations. As a controller becomes more complex, costs increase and initial setting up and calibration becomes more difficult. It is therefore important to select a control system which satisfies the particular needs and process characteristics. A complex PID controller is not necessarily the most suitable for a relatively stable process but could be essential for an unstable process with rapid response times. It could be argued that such a process is not really desirable in water or wastewater treatment in any event!

Further reading

American Water Works Association (1989). *Water Treatment Plant Design*, 2nd edn. Denver: AWWA.
American Water Works Association (1990). *Water Quality and Treatment*, 4th edn. Denver: AWWA,
Chartered Institution of Water and Environmental Management. *Handbooks of UK Wastewater Practice*. London: CIWEM.
   Primary Sedimentation (1973).
   Activated Sludge (1987).
   Biological Filtration (1988).
   Preliminary Processes (1992).
   Tertiary Treatment (1994).
   Sewage Sludge: Stabilization and Disinfection (1996).
   Sewage Sludge: Utilization and Disposal (1996).
   Sewage Sludge: Dewatering, Drying and Incineration (1997).
Dudley, J. and Dickson, C. M. (1994). *Dynamic Modelling of STWs*, UM 1287. Swindon: WRC.
Hall, T. and Hyde, R. A. (1992). *Water Treatment Processes and Practices*. Swindon: WRC.
Metcalf and Eddy Inc. (1990). *Wastewater Engineering: Treatment Disposal Reuse*, 3rd edn. New York: McGraw-Hill.

Montgomery, J. M. (1985). *Water Treatment: Principles and Design*. New York: Wiley.
Rhoades, J. (1997). *An Introduction to Industrial Wastewater Treatment and Control*. London: CIWEM.
Tebbutt, T. H. Y. (1978). Developments in performance relationships for sewage treatment. *Pub. Hlth Engnr*, **6**, 79.
Twort, A. C., Law, F. M., Crowley, F. W. and Ratnayaka, D. D. (1994). *Water Supply*, 4th edn. London: Edward Arnold.
Water Environment Federation (1992). *Design of Municipal Wastewater Treatment Plants*. Alexandria: WEF.
Water Research Centre (1995). *Wastewater and Sludge Treatment Processes*. Swindon: WRC.

10

Preliminary treatment processes

To protect the main units of a treatment plant and to aid in their efficient operation it is necessary to remove the large floating and suspended solids which are often present in the inflow. These materials include leaves, twigs, paper, plastics, rags and other debris which could obstruct flow through a plant or damage equipment in the plant.

10.1 Screening and straining

The first stage in preliminary treatment usually involves a simple screening or straining operation to remove large solids. In the case of water treatment some form of sloping protective boom or coarse screen with openings of about 75 mm is used to prevent large objects reaching the intake. The main screens are usually provided in the form of a mesh with openings of 5–20 mm and arranged as a continuous belt, disc or drum through which the flow must pass (Figure 10.1). The screening mesh is usually rotated slowly so that the material collected can be removed before an excessive head loss is reached The screenings removed from water are normally returned to the source downstream of the abstraction point.

With sewage the content of paper, rags and plastics is often high and the nature of the materials is such that a mesh screen would be difficult to keep clean. It was therefore customary initially to use a bar screen arrangement with a spacing between bars of 10–60 mm. On small works, intermittent hand cleaning of screens is possible but on larger installations automatic mechanical cleaning is provided either on the basis of elapsed time or initiated by the build-up of head loss across the screen. More recently other types and sizes of screens have come into use including moving belt, rotary and drum configurations. Coarse screens are those with apertures >50 mm, medium screens are 15 to 50 mm, fine screens are 3 to 15 mm and milli-screens are defined as having apertures in the range 0.25 to 3 mm. As the aperture size decreases so the speed of blocking and rate of production of screenings increases but unless they can be removed effectively their presence at later stages of the treatment process can be troublesome. As a consequence it has become common to install fine screens at most wastewater treatment facilities and a maximum aperture size of 6 mm is often selected. The

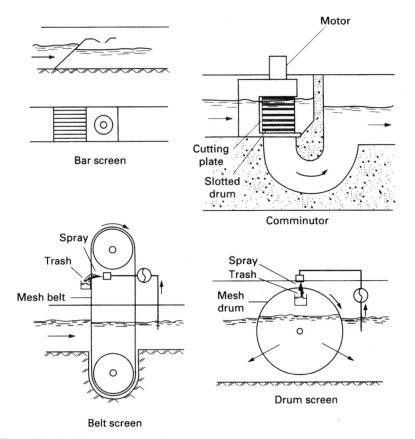

Figure 10.1 Preliminary treatment units.

amount of screenings produced from domestic sewage is variable but is usually in the range of 0.1–0.3 m^3 per 1000 population per hour.

Pollution control authorities frequently require effective screening of CSO discharges and considerable efforts have been made to improve the performance of screens to ensure that aesthetic pollution of watercourses below CSOs is eliminated or reduced. Alternative hydrodynamic solids separators have gained considerable popularity in some areas because their performance can be more reliable than that of screens. In attempts to reduce the aesthetic pollution problem publicity campaigns have been instituted to encourage a 'bag and bin it' approach to the plastic and sanitary items which are frequently disposed of via the WC toilet.

Sewage screenings are unpleasant in nature and are usually disposed of by burial or incineration after any faecal matter has been washed back into the flow. Alternatively, screenings may be passed to a macerator which shreds them to a small size so that they can then be returned to the flow for removal with the rest of the settleable solids during the main treatment process. In some situations the use of a comminutor which shreds the solids *in situ* may be preferred to screening (Figure 10.1) although operational problems with blockages and reconstitution of long strings of debris have greatly reduced their popularity in recent years.

10.2 Microstraining

The microstrainer is a development of the drum screen which uses a fine woven stainless-steel mesh with aperture sizes of 20–40 μm to provide removal of relatively small solids. It has applications in water treatment for removal of algae and similar-size particles from waters of otherwise good quality. Microstraining is also employed as a final tertiary stage to produce a high-quality sewage effluent. Because of the small mesh apertures, clogging occurs rapidly so that the drum is rotated at a peripheral speed of about 0.5 m/s and the mesh continually washed clean by high-pressure sprays. Straining rates in normal usage are 750–2500 m^3/m^2 day. The design of microstrainer installations is based on the laboratory determination of an empirical characteristic of the suspension which measures the behaviour of the suspension with reference to its clogging properties. The results of this test can be used to determine the allowable straining rate to prevent excessive clogging and possible physical damage to the mesh.

10.3 Grit removal

In most sewerage systems and particularly those with combined sewers, considerable amounts of grit are carried along in the flow and this material, if not removed, could cause damage to mechanical parts of the treatment plant. Because the grit particles are relatively large, with a high density compared with the organic particles in sewage, they are often removed using the principle of differential settling. Grit particles with a diameter of 0.20 mm and relative density (specific gravity) of 2.65 have a settling velocity of about 1.2 m/min whereas most of the suspended solids in sewage have considerably lower settling velocities. By using a parabolic section channel it is possible to provide a constant horizontal velocity of around 0.3 m/s at all rates of flow. Under these conditions a channel of sufficient length to provide a retention time of 30–60 seconds will allow the grit particles to settle to the bottom whilst the remaining suspended solids are still transported by the flow. The grit is removed at intervals, washed and then disposed of for re-use in some way. Other types of grit-removal device may involve an aerated spiral-flow chamber to achieve the desired separation or the use of a short retention settling tank, any organic solids removed with the grit being washed back into the flow before the grit is discharged. Grit production is extremely variable and, depending upon the catchment character-istics, will probably be in the range $0.005–0.05\,\mathrm{m^3/1000\,m^3}$ of wastewater.

10.4 Flow measurement and distribution

In order to operate a treatment plant efficiently it is necessary to be able to measure flows into the plant and through the various units. In open channel systems this is usually achieved using an hydraulic control structure, such as a venturi flume, which has a well established head–discharge relationship. For closed-pipe systems a venturimeter, orifice plate or electromagnetic flow meter is used as appropriate.

Flow division is of great importance when treatment capacity is split between several units in parallel. The principle of hydraulic similarity is difficult to achieve and many treatment plants demonstrate the undesirable effects of poor arrangements for flow division. The most effective form of flow splitting is that provided by free-fall weirs arranged in a chamber with a separate length of weir discharging to each unit to be fed. At sewage treatment plants on combined sewerage systems it is necessary to provide storm overflows to spill flows in excess of the flow to full treatment into storm water tanks. This is usually achieved by single or double side weirs which restrict the flow passing on to full treatment to the design value. The excess flow is diverted to the storm tanks which can be pumped out for full treatment when the inflow to the works is no longer affected by surface runoff. The choice of flow measuring and splitting devices on a treatment plant can be restricted by their head loss characteristics since gravity flow through the whole plant is desirable and thus the head available for hydraulic control structures is often limited.

Further reading

Boucher, P. L. (1961). Micro-straining. *J. Instn Pub. Hlth Engnrs*, **60**, 294.

Institution of Water and Environmental Management (1992). *Preliminary Processes*. London: IWEM.

Meeds, B. and Balmforth, D. J. (1995). Full-scale testing of mechanically raked bar screens. *J. C. Instn Wat. Envir. Managt*, **9**, 614.

Roebuck, I. H. and Graham, N. J. D. (1980). Pilot plant studies of fine screening in raw sewage. *Pub. Hlth Engnr*, **8**, 154.

Thomas, D. K., Brown, S. J. and Harrington, D. W. (1989). Screening at marine outfall headworks. *J. Instn Wat. Envir. Managt*, **3**, 533.

White, J. B. (1982) Aspects of the hydraulic design of sewage treatment works. *Pub. Hlth Engnr*, **10**, 164.

11

Clarification

Many of the impurities in water and wastewater occur as suspended matter which remains in suspension in flowing liquids but which will move vertically under the influence of gravity in quiescent or semi-quiescent conditions. Usually the particles are denser than the surrounding liquid so that sedimentation takes place but with very small particles and with low-density particles flotation may offer a more satisfactory clarification process. Sedimentation units have a dual role – the removal of settleable solids and the concentration of the removed solids into a smaller volume of sludge.

11.1 Theory of sedimentation

In sedimentation it is necessary to differentiate between discrete particles which do not change in size, shape or mass during settling and flocculent particles which agglomerate during settling and thus do not have constant characteristics.

The basic theory of sedimentation assumes the presence of discrete particles. When such a particle is placed in a liquid of lower density it will accelerate until a limiting terminal velocity is reached, then

$$\text{gravitational force} = \text{frictional drag force} \tag{11.1}$$

Now

$$\text{gravitational force} = (\rho_s - \rho_w)gV \tag{11.2}$$

where ρ_s = density of particle, ρ_w = density of fluid and V = volume of particle

By dimensional analysis it can be shown that

$$\text{frictional drag force} = C_D A_c \rho_w \frac{v_s^2}{2} \tag{11.3}$$

where C_D = Newton's drag coefficient, A_c = cross-sectional area of particle and v_s = settling velocity of particle.

C_D is not constant, but varies with Reynolds number (R) and, to a lesser extent, with the shape of the particle. For spheres

$$R \leq 1 \qquad C_D = \frac{24}{R} \qquad (11.4)$$

$$1 \leq R \leq 10^3 \qquad C_D = \frac{18.5}{R^{0.6}} \qquad (11.5)$$

$$R > 10^3 \qquad C_D = \frac{24}{R} + \frac{3}{\sqrt{R}} + 0.34 \qquad (11.6)$$

In sedimentation

$$R = \frac{v_s d}{\upsilon} \qquad (11.7)$$

where d = particle diameter and υ = kinematic viscosity of the fluid.

Equating gravitational and frictional forces for the equilibrium condition when the terminal velocity is reached

$$(\rho_s - \rho_w)gV = C_D A_c \rho_w \frac{v_s^2}{2} \qquad (11.8)$$

that is

$$v_s = \sqrt{\frac{2gV(\rho_s - \rho_w)}{C_D A_c \rho_w}} \qquad (11.9)$$

for spheres

$$V = \frac{\pi d^3}{6} \qquad A_c = \frac{\pi d^2}{4}$$

hence

$$v_s = \sqrt{\frac{4gd(\rho_s - \rho_w)}{3 C_D \rho_w}} \qquad (11.10)$$

or

$$v_s = \sqrt{\frac{4gd(S_s - 1)}{3 C_D}} \qquad (11.11)$$

where S_s = specific gravity (relative density) of particle.

For turbulent flow, $10^3 < R$, C_D tends to a value of 0.4. Thus

$$v_s = \sqrt{3.3gd(S_s - 1)} \tag{11.12}$$

For laminar flow, $R \leq 1$, $C_D = 24/R$. Thus

$$v_s = \frac{gd^2(S_s - 1)}{18\upsilon} \tag{11.13}$$

which is Stokes' law.

In calculating settling velocities it is essential to check that the correct formula (11.12) or (11.13) has been used for the velocity as determined. In the transitional range between turbulent and laminar flow a trial and error solution for v_s must be used. Since viscosity is a function of temperature it is important to appreciate that at low temperatures the increased viscosity will reduce settling velocity whereas at high temperatures the settling velocity will increase because of the reduced viscosity. In hot climates the effect of direct sunlight on sedimentation tanks can be such as to create convection currents which exceed in magnitude the temperature-enhanced settling velocities. This is particularly common with relatively light floc suspensions in water treatment where rising sludge can seriously reduce the settling performance. Covering the tank with a light roof can often reduce the effect of direct sun on its performance.

Worked example for settling velocity

A discrete spherical particle has a diameter of 0.15 mm and a relative density of 1.1. Calculate the settling velocity in water at 20°C. (Kinematic viscosity of water at 20°C is $1.01 \times 10^{-6}\,\text{m}^2/\text{s}$.)

Assume that flow is laminar and thus equation 11.13 is appropriate

$$v_s = [9.81 \times (1.5 \times 10^{-4})^2 \times (1.1 - 1.0)]/18 \times 1.01 \times 10^{-6}$$

$$= 0.0012\,\text{m/s}$$

It is now necessary to check that this calculated settling velocity produces a value of R within the laminar range which is assumed in using equation 11.13. From equation 11.7

$$R = 0.0012 \times 1.5 \times 10^{-4}/1.01 \times 10^{-6}$$

$$= 0.178 \text{ which is } <1 \text{ and thus equation (11.13) applies.}$$

If the calculated value of R was >1 it would have been necessary to recalculate v_s using equation 11.5 for the transition range value of C_D and equation 11.11 to determine the new v_s.

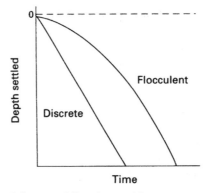

Figure 11.1 Settlement of discrete and flocculent particles.

When dealing with flocculent suspensions it is not possible to apply the above theory because the agglomeration of floc particles results in increasing settling velocity with depth due to the formation of larger and heavier particles; this feature is illustrated in Figure 11.1. Many of the suspensions in the treatment of water and wastewater are flocculent in nature.

Four different types of settling can occur

- class 1 settling: settlement of discrete particles in accordance with theory
- class 2 settling: settlement of flocculent particles exhibiting increased velocity during the process
- zone settling: at certain concentrations of flocculent particles the particles are close enough together for the interparticulate forces to hold the particles fixed relative to one another so that the suspension settles as a unit
- compressive settling: at high concentrations the particles are in contact and the weight of the particles is in part supported by the lower layers of solids.

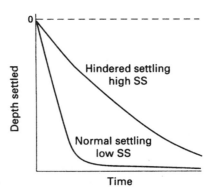

Figure 11.2 Hindered settling.

In the case of concentrated suspensions (>2000 mg/l SS) hindered settlement occurs. In these circumstances there is a significant upward displacement of water due to the settling particles and this has the effect of reducing the apparent settling velocity of the particles (Figure 11.2).

11.2 The ideal sedimentation basin

The behaviour of a sedimentation tank operating on a continuous flow basis with a discrete suspension of particles can be examined by reference to an ideal sedimentation basin (Figure 11.3) which assumes

- quiescent conditions in the settling zone
- uniform flow across the settling zone
- uniform solids concentration as flow enters the settling zone
- solids entering the sludge zone are not resuspended.

Consider a discrete particle with a settling velocity v_0 which just enters the sludge zone at the end of the tank. This particle falls through a depth h_0 in the retention time of the tank t_0, so

$$v_0 = \frac{h_0}{t_0} \tag{11.14}$$

But since, t_0 = Volume/Flow per unit time = V/Q,

$$v_0 = \frac{h_0 Q}{V} = \frac{h_0 Q}{A h_0} \tag{11.15}$$

where A = surface area of tank, then

$$v_0 = \frac{Q}{A} \tag{11.16}$$

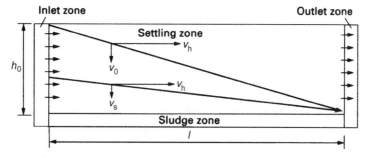

Figure 11.3 The ideal sedimentation basin.

Q/A is termed the surface overflow rate and it follows from equation 11.16 that for discrete particles solids removal is not dependent on the depth of the tank. For flocculent particles, however, depth does affect solids removal since the deeper the tank the more likely it is that agglomeration will occur and hence a larger proportion of the solids would be removed.

If a tank is fed with a suspension of discrete particles of varying sizes it is possible to determine the overall removal as follows, again referring to Figure 11.3.

The tank is designed to remove all particles with settling velocity less than or equal to v_0. Particles with settling velocity $v_s < v_0$ will only be removed if they enter the tank at a distance from the bottom not greater than h where $h = v_s t_0$. Thus the proportion of particles with $v_s < v_0$ which will be removed is given by the ratio v_s/v_0.

By inserting a series of false bottoms in a tank at a spacing of v_s to it would theoretically be possible to remove all solids with a settling velocity of v_s. Thus it should be possible to remove suspended solids with a very low settling velocity provided that the false bottoms or trays were spaced closely enough. Although some use has been made of tanks with an intermediate floor, particularly in water treatment with light hydroxide flocs, the use of multiple floors poses serious problems of sludge removal. In practice what tends to happen is that deposited solids build up on the floors and so restrict the area for flow that the deposits are resuspended by the increased horizontal velocity and the removal efficiency falls. High-rate tube or plate settlers overcome the sludge removal problem by using inclined surfaces which permit deposited sludge to discharge continuously from the bottom of the system. High-rate settlers can be purpose designed but prefabricated units can be inserted into existing conventional settling basins to improve their performance. Inclined tube or plate settlers provide a greatly increased surface area for settlement within the area of the containing tank. The critical settling velocity for discrete particles in a tube or plate settler, as shown diagrammatically in Figure 11.4, is given by

$$v_0 = \frac{kv}{(\sin \alpha + L \cos \alpha)} \tag{11.17}$$

where k = 1.33 for circular tubes and 1.00 for flat plates,
$\quad\quad v$ = velocity of flow through settler elements,
$\quad\quad \alpha$ = angle of inclination of elements and
$\quad\quad L$ = length of element/diameter of tube or distance between plates.

Worked example on ideal settling

A settling basin is designed to be capable theoretically of removing all spherical discrete particles of diameter 0.2 mm and relative density 1.01 from water at

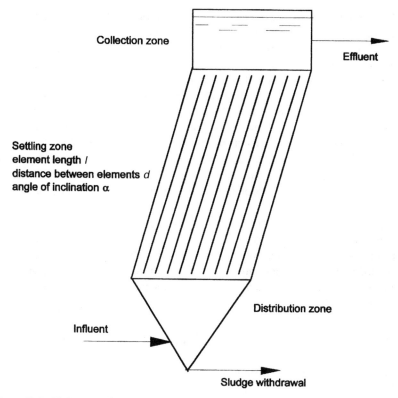

Collection zone

Effluent

Settling zone
element length *l*
distance between elements *d*
angle of inclination α

Distribution zone

Influent

Sludge withdrawal

Figure 11.4 High-rate settler.

20°C. Determine the basin's theoretical removal capability for spherical discrete particles of 0.1 mm diameter and relative density 1.03.

From equation 11.13,

$$v_0 = 9.81 \times (2 \times 10^{-4})^2 \times (1.01 - 1.00)/18 \times 1.01 \times 10^{-6}$$
$$= 0.0021 \text{ m/s}$$

For second particle size

$$v_s = 9.81 \times (1 \times 10^{-4})^2 \times (1.03 - 1.00)/18 \times 1.01 \times 10^{-6}$$
$$= 0.0016 \text{ m/s}$$

Hence removal of second particle size = $0.0016 \times 100/0.0021 = 77.08$ per cent.

11.3 Measurement of settling characteristics

The settling velocity of individual particles may be determined by timing their fall through a known depth of fluid, but for graded suspensions a settling column analysis is more useful. A settling column is a tube 2–3 m deep with tapping points at intervals, the diameter of the tube being at least a hundred times the largest particle size to prevent wall effects.

For an analysis the suspension is thoroughly mixed in the column and the initial SS concentration determined as c_0 mg/l. A sample is taken at depth h_1 after time t_1 and the SS concentration is found to be c_1 mg/l. Now all particles with a settling velocity greater than v_1 ($= h_1/t_1$) will have settled past the sampling point and the particles remaining, i.e. c_1, must have a settling velocity less than v_1. Thus the proportion of particles p_1 having a settling velocity less than v_1 is given by

$$p_1 = \frac{c_1}{c_0} \tag{11.18}$$

The procedure is repeated for time intervals $t_2, t_3, \ldots$, and hence proportions of particles $p_2, p_3, \ldots$ having settling velocities less than $v_2, v_3, \ldots$ are determined. Plotting these data gives the settling characteristic curve for the suspension (Figure 11.5).

In a tank with an overflow velocity of v all particles with $v_s > v$ will be removed regardless of the position at which they enter the tank. In addition, for a horizontal flow tank, particles with $v_s < v$ will be removed if they enter at a

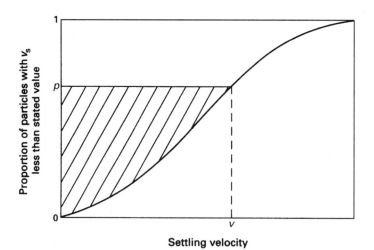

Figure 11.5 Settling characteristic curve for a discrete suspension.

distance from the bottom not exceeding $v_s t_0$. Thus from Figure 11.5 the overall removal in a horizontal flow tank is given by

$$P = 1 - p + \int_0^p \frac{v_s}{v} \, dp \qquad (11.19)$$

For a vertical flow tank

$$P = 1 - p \qquad (11.20)$$

since all particles with $v_s < v$ will eventually be washed out of the tank. Figure 11.6 gives a diagrammatic illustration of this point.

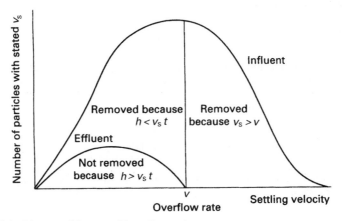

Figure 11.6   Discrete solids removal by sedimentation.

When dealing with flocculent suspensions both the concentration of particles and the effect of depth of flocculation must be taken into account. Samples are taken from a number of depths at each time interval and the SS in each sample are expressed as a percentage of the original concentration. The difference between this percentage and 100 is thus the percentage of solids which have settled past the sampling point and would therefore have been removed in a tank of that particular depth and retention time. Plotting these results as in Figure 11.7 enables the construction of smooth isoconcentration lines for the SS removal at various depth and time conditions. The effect of flocculation is shown in the shape of the isoconcentration lines, the greater the slopes of the lines with depth the greater the degree of flocculation which has taken place. The performance of a specific tank and flocculent suspension combination can be estimated by taking the removal indicated by the isoconcentration plot for the tank depth and retention time. Additional removals of smaller particles which enter the tank at

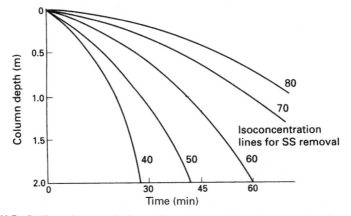

Figure 11.7 Settling column results from a flocculent suspension. Isoconcentration lines for SS removal drawn around the experimental results for a range of depths and sampling times.

a suitable height to be trapped are obtained by taking bands of concentrations and multiplying the band width by the ratio of the mid point of that band to the tank depth.

Worked example on flocculent settling

The following results were obtained from a settling column analysis on a flocculent suspension

Time (min)	0	20	40	60	80
Depth (m)	SS concentration (mg/l)				
0.5	480	216	120	48	24
1.0	480	288	197	115	43
2.0	480	370	274	187	120
3.0	480	384	302	230	158

The SS concentrations in the above table are converted into percentages of the original SS concentration of 480 mg/l and the percentage removals at the various times and depths calculated with the results shown in the table on the following page.

Time (min)	20	40	60	80
Depth (m)	SS removal (%)			
0.5	55	75	90	95
1.0	40	59	76	91
2.0	23	43	61	75
3.0	20	37	52	67

These removals are then plotted as isoconcentration lines (approximated to straight lines) in Figure 11.8 which can then be used to determine the performance of a settling tank with a surface overflow rate of 48 m/day and a

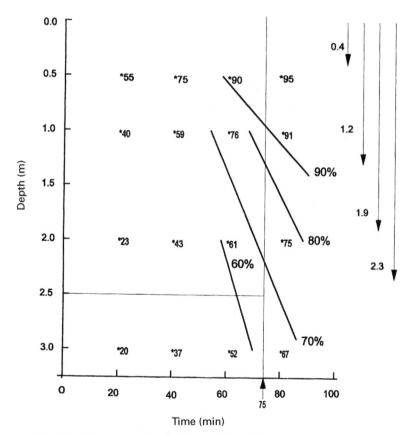

Figure 11.8 Use of isoconcentration plot for flocculent settlement.

depth of 2.5 m which corresponds to a retention time of 1.25 h (75 min). Inspection of the plot in Figure 11.8 indicates a removal of 67 per cent at the co-ordinates of 75 min and 2.5 m. Estimates of the mid points of the additional percentage removal bands give the total removal as

$$\text{Total removal} = 67 + (1/2.5) \left[(70 - 67)2.3 + (80 - 70)1.9 + (90 - 80)1.2 + (100 - 90)0.4\right]$$

$$= 84 \text{ per cent}$$

11.4 Efficiency of sedimentation tanks

The hydraulic behaviour of a tank may be examined by injecting a tracer into the inlet and observing its appearance in the effluent. The flow-through curves so obtained are of infinite variety, ranging from the ideal plug-flow case to that of a completely mixed tank as shown in Figure 11.9. The flow-through curve

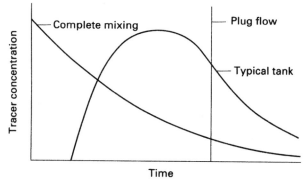

Figure 11.9 Flow-through curves.

obtained in practice is a combination of the two extremes, short-circuiting due to density currents and mixing due to hydraulic turbulence producing a peak earlier than would be expected in an ideal tank. Thus the actual retention time is often considerably less than the theoretical value. The residence time characteristics of any flow-through reactor are important as they can have a major influence on its performance.

Since the purpose of sedimentation tanks is to remove suspended matter the logical way of expressing their efficiency is by the percentage removal of such solids. The normal SS determination records particles down to a few microns in size whereas floc particles smaller than 100 μm are unlikely to be removed by sedimentation. Thus a sedimentation tank will never remove all the SS from sewage and the normal range of SS removal from sewage by sedimentation is 50–70 per cent. Research has shown that with heterogeneous suspensions such as sewage the hydraulic loading on a tank has less influence on the removal efficiency than the influent SS concentration. In most cases the higher the initial SS concentration the greater will be the percentage removal.

11.5 Types of sedimentation tank

The main conventional types of sedimentation tank found in practice are shown in Figure 11.10. The horizontal tank is compact but suffers from a restricted effluent

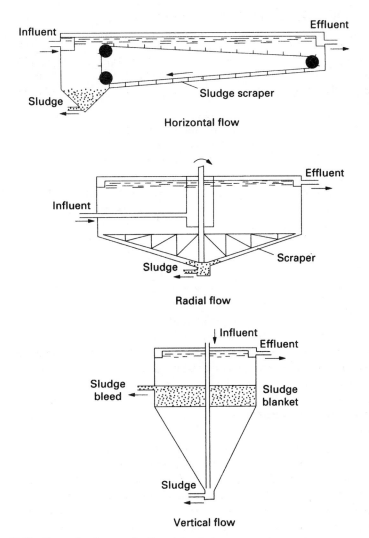

Figure 11.10   Conventional types of sedimentation tank.

weir length unless suspended weirs are adopted. Sludge is moved to the sump by a travelling bridge scraper which may serve several tanks or by a continuous-belt system with flights. The sludge is withdrawn from the sump under hydrostatic head. Circular tanks offer advantages of long weir length and simpler scraping mechanisms but are not so compact. Hopper-bottom tanks with horizontal flow are popular on small sewage works where the extra construction cost is more than offset by the absence of any scraping mechanism. Rectangular horizontal-flow tanks are most efficient in utilization of land area but the sludge scraping mechanisms tend to be somewhat troublesome and the weir length available for effluent discharge is limited. Circular tanks are usually more stable hydraulically, have more reliable sludge scrapers and longer effluent weir lengths. They are, however, more costly to construct than rectangular tanks. The vertical-flow hopper-bottom tank is often used in water-treatment plants and in such conditions operates with a sludge blanket which serves to strain out particles smaller than would be removed by sedimentation alone at the overflow rate employed. Because of the high construction cost of hopper-bottom tanks many vertical-flow units now have flat bottoms with complex inlet distributors to provide the even flow characteristics which are inherent in the inverted pyramid configurations.

Many different designs of inlet and outlet structures are in use and whilst particular designs may offer some improvement in solids removal with homogeneous flocculent suspensions they usually make little difference to the removal of SS from raw sewage. It is, however, important to keep velocities as low as possible in the vicinity of effluent weirs to avoid entrainment of suspended solids. Inset weirs can be used in rectangular tanks to increase available length and with all effluent weirs it is good practice to use slotted vee-notch weir plates which can be accurately levelled and which are in any case relatively insensitive to slight variations in level.

Sedimentation tanks have two functions: the removal of settleable solids to produce an acceptable output and the concentration of the removed solids into a smaller volume. The design of a tank must consider both of these functions and the tank should be sized on whichever of the requirements is limiting. The sludge thickening function of a tank is likely to be important when dealing with relatively high concentrations of homogeneous solids.

11.6 Gravity thickening

The design of sedimentation tanks must take into account both the solids removal and the sludge-thickening functions. The size of a unit will be limited by one of these functions and in the case of high concentrations of homogeneous suspensions like activated-sludge flocs or chemical precipitation flocs the thickening function may be the more critical.

Analogous to the ideal settling-basin concept it is possible to envisage an ideal thickener which has uniform horizontal distribution of particles and from the base of which the thickened suspension is removed creating equal downward velocity across the tank section. It is also assumed that an ideal suspension having incompressible solids is present. As shown in Figure 11.11 the flux of solids past a point in the thickener is the result of the rate of downward movement under gravity and the rate of downward movement due to the removal of solids by withdrawal of sludge. The solid flux in a continuous thickener, G_c is given by

$$G_c = c_i v_i + c_i v_w \qquad (11.21)$$

where c_i = solids concentration, v_i = settling velocity of solids at concentration c_i and v_w = downward velocity produced by withdrawal.

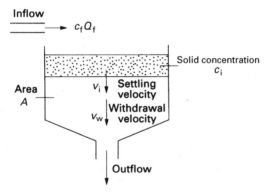

Figure 11.11 Gravity thickening.

The batch settlement flux $(c_i v_i)$ is governed by the physical characteristics of the suspension. The withdrawal flux $(c_i v_w)$ is an operational parameter directly proportional to the sludge-removal rate. Figure 11.12 shows how the two fluxes change as the solids concentration increases. The reason for the shape of the settling flux plot is that at low SS levels the settling velocity will be high but as the SS level increases hindered settling effects cause a progressive reduction in the bulk settling velocity. The total flux curve can then be used to aid in the design of the thickening function of clarifiers. Considering the suspension whose characteristics are shown in the lower portion of Figure 11.12, it is clear that the concentration c_1 corresponds to a minimum total flux level so that to thicken the sludge to a concentration of c_1 or higher it will be necessary to ensure that the applied solids load does not exceed G_1, that is

$$\text{Applied load} = \frac{c_f Q_f}{A} \le G_1 \qquad (11.22)$$

where c_f = solids concentration in tank feed, Q_f = flow rate to tank and A = cross-sectional area of tank.

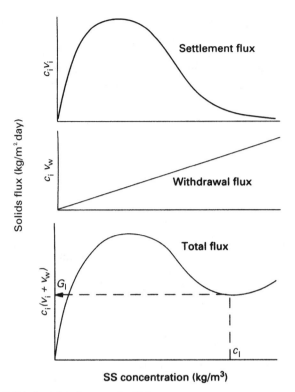

Figure 11.12   Solids flux plots for a gravity thickener.

If the solids flux is higher than G_1 not all of the solids would be able to reach the sludge outlet. To remove all solids at a flux greater than G_1, it would be necessary either to increase the area of the tank or to increase the removal velocity which would give reduced thickening since the solids concentration in the withdrawn flow is inversely proportional to the removal velocity.

It is possible to obtain this type of information about the thickening characteristics of suspensions directly from the batch settlement flux curve alone. It can be shown that if a tangent is drawn to the settlement flux curve at the rate-limiting concentration the slope of the tangent will be the necessary downward velocity in a continuous thickener to maintain a constant sludge level. The intercept on the y-axis will be the solids flux for the equilibrium condition and the intercept on the x-axis will be the solids concentration in the withdrawn sludge. Figure 11.13 shows the use of the settlement curve in this manner. The slope of the line from the origin to the point of tangency is the gravitational settling velocity at c_1 and since the slope of the tangent is v_w, the intercept is G_1 and the total flux made up from the two components as shown. This type of plot may be used to determine the required thickener area for various thickened sludge

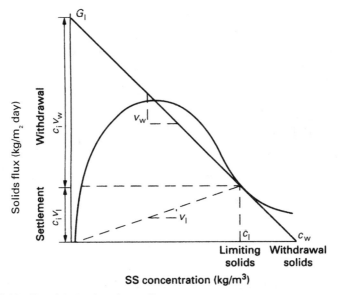

Figure 11.13 Use of the batch settlement flux curve.

concentrations. Similarly the plot can be used to predict the effect of a change in loading on an existing tank with respect to sludge concentration and the necessary withdrawal rate.

11.7 Flotation

Some of the suspensions encountered in water and wastewater treatment are composed of small size, low density particles which have relatively low settling velocities even when treated by chemical coagulation as described in the next chapter. An alternative clarification technique which is particularly attractive for such suspensions is to encourage the particles to float to the surface where they can be removed as a scum. Some suspensions have a density so close to that of the fluid in which they are suspended that they need little encouragement to float. Even with particles whose density is greater than that of the suspending fluid it is possible to use flotation by the addition of an agent which produces positive buoyancy. Air bubbles are effective flotation agents and there are a number of techniques which use the release or generation of air bubbles to provide the necessary buoyancy. By far the most popular technique is that of dissolved air flotation (DAF) which is shown diagrammatically in Figure 11.14. The process involves saturation of a recycled portion of the throughput (usually about 10 per cent) through a saturator where air is injected at high pressures of up to 400 kPa. This pressurized recycle flow is returned to the inlet at the bottom of the flotation tank where it is mixed with the incoming flow. The sudden drop in pressure

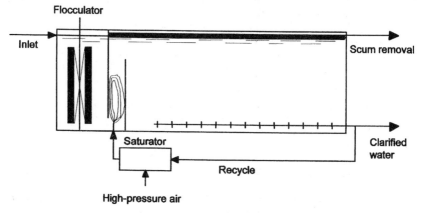

Figure 11.14 Dissolved air flotation.

produces supersaturation of the water with air and releases a cloud of fine air bubbles which attach themselves to particles in the suspension and cause them to float. Dissolved air flotation has particular advantages in water treatment when chemical coagulation is used to remove dissolved substances such as colour, iron and manganese. In such circumstances rise rates of up to 12 m/h can be achieved as compared with normal sedimentation rates of perhaps 4 m/h. This means that flotation units are usually much smaller than the alternative sedimentation tanks and have the advantage of producing a lower turbidity output. Dissolved air flotation units can be brought into full operation in a short space of time whereas conventional sedimentation tanks often take several days to reach equilibrium operating conditions. The scum removed from flotation units is normally of a higher solids content than would be produced by settlement of the same suspension so that the volume of sludge is significantly reduced. The capital cost of flotation units is less than that of the equivalent settling facility but operating costs are higher because of the power required for pumping and compressed air injection. Flotation units tend to require closer control than sedimentation units because their performance can be upset by relatively small changes in raw water quality or operational parameters. The total cost of dissolved air flotation will probably be cheaper than conventional sedimentation for suspensions where a settling tank overflow rate >2 m/h cannot be used.

Further reading

Dick, R. I. (1972). Gravity thickening of sewage sludges. *Wat. Pollut. Control*, **71**, 368.

Fadel, A. A. and Baumann, R. E. (1990). Tube settler modelling. *J. Envir. Engng Am. Soc. Civ. Engnrs*, **116**, 107.

Hazen, A. (1904). On sedimentation. *Trans. Am. Soc. Civ. Engnrs*, **53**, 45. (Reprinted *J. Proc. Inst. Sew. Purif.*, 1961, 521(6).)

Institution of Water and Environmental Management (1972). *Primary Sedimentation.* London: IWEM.

Kynch, G. J. (1952). A theory of sedimentation. *Trans. Faraday Soc.*, **48**, 166.

Malley, J. R. and Edzwald, J. K. (1991). Laboratory comparison of DAF with conventional treatment. *J. Amer. Wat. Wks Assn*, **83**(9), 56.

Richard, Y. and Capon, B. (1980). Sedimentation. In *Developments in Water Treatment*, Vol. 1 (W. M. Lewis, ed.). Barking: Applied Science Publishers.

Tebbutt, T. H. Y. (1979). Primary sedimentation of wastewater. *J. Wat. Pollut. Control Fedn*, **51**, 2858.

Tebbutt, T. H. Y. and Christoulas, D. G. (1975). Performance relationships for the primary sedimentation. *Wat. Res.*, **9**, 347.

Willis, R. M. (1978). Tubular settlers – a technical review. *J. Am. Wat. Wks Assn*, **70**, 331.

Zabel, T. F. and Melbourne, J. D. (1980). Flotation. In *Developments in Water Treatment*, Vol. 1 (W. M. Lewis, ed.). Barking: Applied Science Publishers.

Problems

1. Find the settling velocity of spherical discrete particles 0.06 mm diameter, relative density 2.5 in water at 20°C ($v = 1.010 \times 10^{-6}\,m^2/s$). (0.0029 m/s)

2. A settling tank is designed to remove spherical discrete particles 0.5 mm diameter, relative density 1.01 from water at 20°C. Assuming ideal settling conditions, determine the removal of spherical discrete particles 0.2 mm diameter, relative density 1.01 by this tank. (16 per cent)

3. Settling column tests on a discrete particle suspension gave the following results from a depth of 1.3 m.

Sampling time (min)	5	10	20	40	60	80
% of initial SS in sample	56	48	37	19	5	2

Determine the theoretical removal of solids from this suspension in a horizontal flow tank with surface overflow rate of 200 m^3/m^2 day. (67 per cent)

4. Tests on a flocculent suspension in a settling column with three sampling points gave the following results.

Sample time (min)	% SS removed at		
	1 m	2 m	3 m
0	0	0	0
10	30	18	16
20	60	48	40
30	62	61	60
40	70	63	61
60	73	69	65

Estimate the probable removal of solids from this suspension in a tank 2 m deep with retention time of 25 min. (64 per cent)

5. Laboratory studies on a floc suspension produced the following settling data.

SS (mg/l)	2500	5000	7500	10 000	12 500	15 000	17 500	20 000
Settling velocity (mm/s)	0.80	0.41	0.22	0.10	0.04	0.02	0.01	0.01

If the suspension is to be thickened to a concentration of 2 per cent (20 000 mg/l) determine the thickener cross-sectional area required for a flow of 5000 m³/day with an initial SS of 3000 mg/l. If a thickened sludge solids content of 1.5 per cent was acceptable determine the new cross-sectional area required. Determine the minimum cross-sectional area required for the settling function of the tank if the settling velocity is assumed to be 0.8 mm/s in this region of the tank. (148 m², 63 m², 72.3 m²)

12

Coagulation

Many impurities in water and wastewater are present as colloidal solids which will not readily settle. Their removal can, however, often be achieved by promoting agglomeration of such particles by flocculation, with or without the use of a coagulant followed by sedimentation or flotation.

12.1 Colloidal suspensions

Sedimentation can be used to remove suspended particles down to a size of about 50 μm depending on their density, but smaller particles have very low settling velocities so that removal by sedimentation is not feasible. Table 12.1 gives calculated settling velocities for particles with relative density 2.65 in water at 10°C. It can be seen that in practical terms the smaller particles have virtually non-existent settling velocities. If these colloidal particles can be persuaded to agglomerate they may eventually increase in size to such a point that removal by sedimentation becomes possible. In a quiescent liquid tiny particles collide because of Brownian movement and collisions also occur when rapidly settling solids overtake more slowly settling particles. As a result larger particles, fewer in number, are produced; growth by these means is, however, slow. Collisions between particles can be improved by gentle agitation, the process of flocculation, which may be sufficient to produce settleable solids from a high concentration of colloidal particles. With low concentrations of colloids a coagulant is added to produce bulky floc particles which enmesh the colloidal solids.

Table 12.1 Settling velocities for discrete particles of relative density 2.65 in water at 10°C

Particle size (μm)	Settling velocity (m/h)
1000	6×10^2
100	2×10^1
10	3×10^{-1}
1	3×10^{-3}
0.1	1×10^{-5}
0.01	2×10^{-7}

12.2 Flocculation

Agitation of water by hydraulic or mechanical mixing causes velocity gradients the intensity of which controls the degree of flocculation produced. The number of collisions between particles is directly related to the velocity gradient and it is possible to determine the power input required to give a particular degree of flocculation as specified by the velocity gradient.

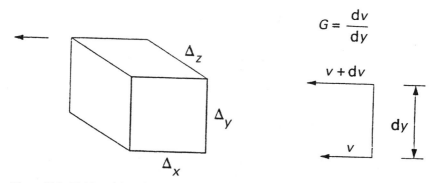

Figure 12.1 Fluid particle undergoing flocculation.

Consider an element of fluid undergoing flocculation (Figure 12.1). The element will be in shear and thus

$$\text{power input} = \tau \Delta_x \Delta_z \Delta_y \frac{dv}{dy} \tag{12.1}$$

where τ = shear stress.

$$\text{Power per unit volume} = P = \tau \frac{\Delta_x \Delta_y \Delta_z}{\Delta_x \Delta_y \Delta_z} \frac{dv}{dy} \tag{12.2}$$

$$= \tau \frac{dv}{dy} \tag{12.3}$$

but by definition, $\tau = \mu \, dv/dy$, where μ = absolute viscosity. Thus

$$P = \mu \frac{dv}{dy} \frac{dv}{dy} = \mu \left(\frac{dv}{dy}\right)^2 \tag{12.4}$$

and putting $G = dv/dy$,

$$P = \mu G^2 \tag{12.5}$$

For hydraulic turbulence in a baffled tank where 'round the end' or 'up and over' flow patterns are created by horizontal or vertical baffles with a minimum spacing of around 0.5 m and a minimum gap between baffle end and wall or water surface of about 0.6 m with velocities of 0.1–0.3 m/s,

$$P = \frac{\rho g h}{t} \tag{12.6}$$

where ρ = mass density of the fluid, h = head loss in tank and t = retention time in tank (usually 15–20 min).

Now from equation 12.5, $G = \sqrt{P/\mu}$, so

$$G = \sqrt{\frac{\rho g h}{\mu t}} = \sqrt{\frac{g h}{\upsilon t}} \tag{12.7}$$

In the case of a mechanically stirred tank

$$P = \frac{Dv}{V} \tag{12.8}$$

where D = drag force on paddles, v = velocity of paddles and V = volume of tank.

From equation 11.3,

$$D = C_D A \rho \frac{v_r^2}{2} \tag{12.9}$$

where v_r = velocity of paddles relative to the fluid in the tank (usually about three-quarters of the paddle velocity v) and A = cross-sectional area of paddles perpendicular to direction of motion.

Thus

$$P = \frac{C_D A \rho v_r^2 v}{2V} \tag{12.10}$$

Earlier editions of this book, in common with most other texts, wrote equation 12.10 as

$$P = \frac{C_D A \rho v_r^3}{V}$$

but this is incorrect. The drag force is certainly a function of the relative velocity, but the power required is the product of the force required to overcome the drag on the paddles and the absolute velocity at which the paddles are moving. The author is grateful to Dr Adrian Coad formerly of WEDC at Loughborough University for bringing this point to his attention.

Substituting for P from equation 12.5

$$G^2 = \frac{C_D A \rho v_r^2 v}{2V\mu} \tag{12.11}$$

that is,

$$G = \sqrt{\frac{C_D A v_r^2 v}{2vV}} \tag{12.12}$$

For good flocculation G should be in the range of 20–70 m/s m. Lower values will probably give inadequate flocculation and higher values will tend to shear the larger floc particles. There is some benefit in providing tapered flocculation with higher G values near the entry and lower values close to the exit. The normal retention time in mechanical flocculation tanks is 20–30 min but there is evidence that within limits the product Gt is important and a typical value of 5 to 10×10^4 is often quoted. With mechanical flocculation the tank depth is usually between one-and-a-half and twice the paddle diameter and the blade area is 10–25 per cent of the tank cross-sectional area.

Mechanical flocculators provide more control over the process than hydraulic flocculators but require more maintenance. In particular, hydraulic flocculators are designed to produce the required velocity gradient at a specific rate of flow. If the flow rate through the unit is significantly altered there will be consequent effects on the flocculation process. This can cause problems if the unit is required to accept a larger flow when a plant is uprated in capacity. The size and speed of mechanical paddles can readily be changed to maintain the required degree of flocculation if the flow is increased. Flocculation and sedimentation may be combined in a single unit (Figure 12.2) and the sludge blanket type of tank is very popular for water treatment purposes.

12.3 Coagulation

Flocculation of dilute colloidal suspensions provides only infrequent collisions and agglomeration does not occur to any marked extent. In such circumstances clarification is best achieved using a chemical coagulant followed by flocculation and sedimentation. Before flocculation can take place it is essential to disperse the coagulant, usually required in doses of 30–70 mg/l, throughout the body of water. This is carried out in a rapid mixing chamber with a high-speed turbine (Figure 12.3) or by adding the coagulant at a point of hydraulic turbulence, e.g. at a hydraulic jump in a measuring flume or at a weir. A velocity gradient of around $1000 \, s^{-1}$ is required for effective mixing. The coagulant is usually a metal

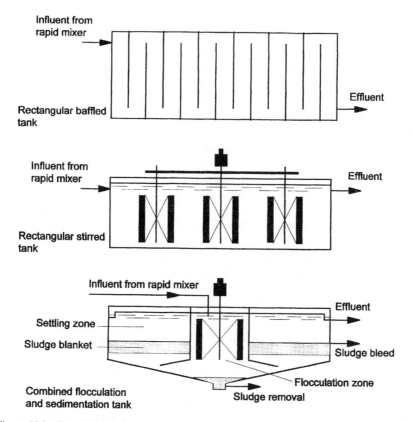

Figure 12.2 Flocculation tanks.

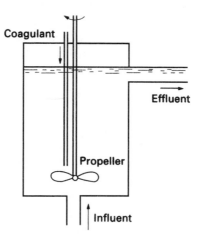

Figure 12.3 Rapid mixer.

salt which reacts with alkalinity in the water to produce an insoluble metal hydroxide floc which incorporates the colloidal particles. This fine precipitate is then flocculated to produce settleable solids. Coagulants are usually added as a concentrated solution which can be dosed accurately using a positive-displacement metering pump.

For many years aluminium sulphate has been the most popular coagulant for water treatment, although as mentioned in Chapter 5 there have been some concerns about possible health hazards from aluminium residuals following its use in drinking water. It is perhaps worth noting that the 1993 WHO *Guidelines for Drinking Water Quality* do not consider aluminium as a health-related constituent. The reactions which take place when aluminium sulphate, known commercially as alum, is added to water are complex and are often simplified as

$$Al_2(SO_4)_3 + 6H_2O \rightarrow \underline{2Al(OH)_3} + 3H_2SO_4$$

$$3H_2SO_4 + 3Ca(HCO_3)_2 \rightarrow 3CaSO_4 + 6H_2CO_3$$

$$6H_2CO_3 \rightarrow 6CO_2 + 6H_2O$$

i.e. overall,

$$Al_2(SO_4)_3 + 3Ca(HCO_3)_2 \rightarrow \underline{2Al(OH)_3} + 3CaSO_4 + 6CO_2$$

When using commercial alum $Al_2(SO_4)_3.14H_2O$ it is found that

> 1 mg/l alum   destroys 0.5 mg/l alkalinity as $CaCO_3$
>   produces 0.44 mg/l carbon dioxide

Thus for satisfactory coagulation sufficient alkalinity must be available to react with the alum and also to leave a suitable residual in the treated water to provide pH buffering.

The solubility of $Al(OH)_3$ is pH dependent and is low between pH 5 and 7.5; outside this range coagulation with aluminium salts is not successful. Other coagulants sometimes used are

- ferrous sulphate (copperas), $FeSO_4.7H_2O$
- ferric sulphate, $Fe_2(SO_4)_3$
- ferric chloride, $FeCl_3$

Copperas is sometimes treated with chlorine to give a mixture of ferric sulphate and ferric chloride known as chlorinated copperas. Ferric salts give satisfactory coagulation above pH 4.5, but ferrous salts are only suitable above pH 9.5. Iron salts are cheaper than alum but unless precipitation is complete residual iron in solution can be troublesome, particularly due to its stain-producing properties in washing machines. Nevertheless, iron salts are now again becoming more popular because of the probably unjustified public concern about aluminium in water. An alternative to conventional coagulation, known as the *Sirofloc* process, uses a suspension of finely divided magnetite to adsorb colour, turbidity, iron and aluminium from water. After contact with the incoming water in the presence of pH-control reagents and a coagulant aid, the flow is passed through a strong magnetic field which causes agglomeration of the magnetite particles. These agglomerates are then removed in an upward flow settling tank and the magnetite sludge recovered for desorption of the contaminants and re-use.

With very low concentrations of colloidal matter floc formation is difficult and coagulant aids may be required. These may be simple additives like clay particles which form nuclei for precipitation of the hydroxide or polyelectrolytes, heavy long-chain synthetic polymers, which added in small amounts (<1 mg/l) promote agglomeration and toughen the floc. Because of the spongy nature of floc particles they have a very large surface area and are thus capable of adsorbing some dissolved organic matter from solution. This surface-active effect, together with chemical reactions between the coagulant and the organic colour results in coagulation removing some dissolved colour as well as colloidal turbidity from water. A sample of raw water with colour 60°H and turbidity 30 NTU would usually be improved to about 5–20°H and 5 NTU after coagulation, flocculation and sedimentation.

A number of natural substances have coagulating properties and some have been used for centuries in developing countries as a means of clarifying water in storage containers. Several of these substances have more recently received attention as low-cost coagulants in treatment plants. Most of these natural substances are derived from the seeds, bark or sap of trees and plants which contain polyelectrolytes. The most commonly used is the *Moringa oleifera* seed which has been used as a primary coagulant in the Sudan and a number of other countries. Its performance with turbid surfacewaters has been shown to be similar to that of conventional coagulants like aluminium sulphate when used in the same concentration. A particular advantage of the *Moringa oleifera* is that the seeds contain 40 per cent edible vegetable oil and after the oil has been extracted the press cake residue still contains the natural cationic polyelectrolyte. The seeds can thus provide both nutrition and an effective coagulant. There have been some concerns that the organic matter present in the seed may promote bacterial growth in the water but if effective disinfection is provided this should not pose a serious problem.

It is not possible to calculate the dose of coagulant required nor the results that it will produce so that laboratory tests must be carried out using the jar-test procedure. This involves setting up a series of samples of water on a special multiple stirrer rig and dosing the samples with a range of coagulant, e.g. 0, 10, 20, 30, 40 and 50 mg/l stirring vigorously. The samples are then flocculated for 30 min and allowed to stand in quiescent conditions for 60 min. The supernatant water is then examined for colour and turbidity and the lowest dose of coagulant to give satisfactory removal is noted. A second set of samples is prepared with pH adjusted over a range, for example 5.0, 6.0, 6.5, 7.0, 7.5, 8.0, and the coagulant dose determined previously added to each beaker followed by stirring, flocculation and settlement as before. It is then possible to examine the supernatant and select the optimum pH and, if necessary, recheck the minimum coagulant dose required. Figure 12.4 shows typical results from such a jar test. Because of the effect of pH on coagulation it is normally necessary in chemical coagulation plants to make provision for the control of pH by the addition of acid or alkali.

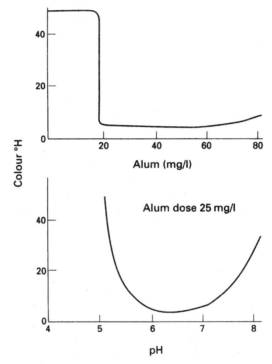

Figure 12.4 Jar-test results.

12.4 Mechanism of coagulation

Although chemical coagulation is a widely used process the mechanisms by which it operates are not fully understood in spite of considerable research effort. Basic colloid stability considerations have been applied to coagulation in attempts to offer explanations for the observed results. The stability of hydrophobic colloid suspensions can be explained by consideration of the forces acting on the particles as shown in Figure 12.5. Mutual repulsion arises from the electrostatic surface charges but destabilization can be achieved by the addition of ions of opposite charge to reduce the repulsive forces and permit the molecular attraction forces to

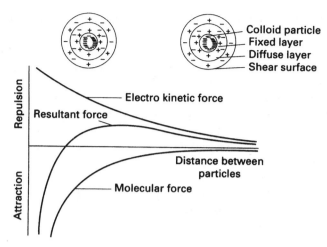

Figure 12.5   Forces acting on a floc particle.

become dominant. In this context the value of the zeta potential, the electrical potential at the edge of the particle agglomerate, is of some significance. In theory a zero zeta potential should provide the best conditions for coagulation. However, when dealing with the heterogeneous suspensions found in water it seems that there are many complicating factors and zeta potential measurements are not always of much value in operational circumstances.

In the case of relatively low suspended solids concentrations, coagulation usually occurs by enmeshment in insoluble hydrolysis products formed as the result of a reaction between the coagulant and the water. In this 'sweep coagulation' the nature of the original suspended matter is of little significance and it is the properties of the hydrolysis product which control the reaction. Unfortunately, the behaviour of coagulants when added to water can be highly complex. In the case of aluminium sulphate the simplified reactions given earlier in this chapter are now known to be far removed from the actual situation. The hydrolysis products of aluminium are very complex, their nature being affected by such factors as the age and strength of the coagulant solution. Hydrolysis products of aluminium include compounds of the form

$$[Al(H_2O)_5OH]^{2+} \quad \text{and} \quad [Al_6(OH)_{15}]^{3+}$$

and sulphate complexes may also appear. As a result the actual reactions taking place are difficult to specify. The situation is further complicated if natural colour is present because the aluminium sulphate reacts with the organics acids responsible for colour.

With higher suspended solids concentrations the colloidal theory can provide a basis for explaining the observed reactions. Thus destabilization of a colloidal suspension occurs due to the adsorption of strongly charged partially hydrolysed metallic ions. Continued adsorption results in charge reversal and restabilization of the suspension which does occur with high coagulant doses. In this type of situation the nature of the colloidal particles does therefore have an influence on the coagulation process.

When coagulation is used to remove colour from water the reaction appears to depend upon the formation of precipitates from the combination of the soluble organics and the coagulant. There is thus generally a direct relationship between colour concentration and the dose of coagulant required for removal of the colour.

The value of coagulant aids can be related to the ability of large molecules in the form of long-chain structures to provide a bridging and binding action between adjacent suspended particles thus promoting agglomeration and preventing floc break-up under shear. With ionic coagulant aids, charge neutralization will also occur as with primary coagulants although at the normal doses employed for coagulant aids this effect is not likely to be very important.

Further reading

Critchley, R. F., Smith, E. O. and Pettit, P. (1990). Automatic coagulation control at water treatment plants in the North West region of England. *J. Instn Wat. Envir. Managt*, **4**, 535.

Folkard, G. K., Sutherland, J. P. and Al-Khalili, R. S. (1996). Natural coagulants – a sustainable approach. In *Sustainability of Water and Sanitation Systems* (J. R. Pickford, ed.). London: Intermediate Technology Publications.

Franklin, B. C., Hudson, J. A., Warnett, W. R. and Wilson, D. (1993). Three-stage treatment of Pennine waters. *J. Instn Wat. Envir. Managt*, **7**, 223.

Gregory, R., Maloney, R. J. and Stockley, M. (1988). Water treatment using magnetite: A study of a Sirofloc pilot plant. *J. Instn Wat. Envir. Managt*, **2**, 532.

Hilson, M. A. and Richards, W. N. (1980). Polymeric flocculants. In *Developments in Water Treatment*, Vol 1 (W. M. Lewis, ed.). Barking: Applied Science Publishers.

Holm, G. P., Stockley, M. and Shaw, G. (1992). The Sirofloc process at Redmires water-treatment works. *J. Instn Wat. Envir. Managt*, **6**, 10.

Ives, K. J. and Bhole, A. G. (1973). Theory of flocculation for continuous flow system. *J. Envr. Engng Div. Am. Soc. Civ. Engnrs*, **99**, 17.

Packham, R. F. and Sheiham, I. (1977). Developments in the theory of coagulation and flocculation. *J. Instn Wat. Engnrs Scientists*, **3**, 96.

Stevenson, D. G. (1980). Coagulation and flocculation. In *Developments in Water Treatment*, Vol. 1 (W. M. Lewis, ed.). Barking: Applied Science Publishers.

Problems

1. A flocculation tank 10 m long, 3 m wide and 3 m deep has a design flow of 0.05 m³/s. Flocculation is achieved by three paddle wheels each with two blades 2.5 m by 0.3 m, the centre line of the blades being 1 m from the shaft which is at mid depth of the tank.

The paddles rotate at 3 rev/min and the velocity of the water is 25 per cent of the blade velocity. For water at 20°C, $v = 1.011 \times 10^{-6} \, m^2/s$. Calculate the power required for flocculation and the velocity gradient. (70.8 W, 23.2 m/s m)

2. A water supply with 15 mg/l alkalinity requires 40 mg/l aluminium sulphate for coagulation. Calculate the quantity of hydrated lime $Ca(OH)_2$ required to leave a finished water with 25 mg/l alkalinity. How much soda ash, Na_2CO_3, would be needed if it were used in place of lime? Ca 40, O 16, H 1, Na 23. ($22.2 \, g/m^3$, $31.8 \, g/m^3$)

13

Flow through porous media

Filtration of suspensions through porous media, usually sand, is an important stage of the treatment of potable waters to achieve final clarity. Although about 90 per cent of the turbidity and colour are removed in coagulation and sedimentation a certain amount of floc is carried over from settling tanks and requires removal. Sand filtration is also employed to provide tertiary treatment of 30:20 standard sewage effluents. Other uses of flow through porous media include ion-exchange beds, adsorption beds and absorption columns where the aim is not to remove suspended matter but to provide contact between two systems.

13.1 Hydraulics of filtration

The resistance to flow of liquids through a porous medium is analogous to flow through small pipes and to the resistance offered by a fluid to settling particles.

The basic formulae for the hydraulics of filtration assume a bed of uni-size medium and refer to the schematic filter shown in Figure 13.1.

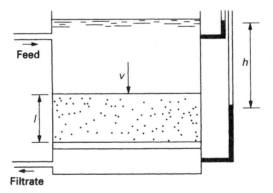

Figure 13.1 Schematic filter bed.

The earliest filtration formula due to Darcy is

$$\frac{h}{l} = \frac{v}{k} \tag{13.1}$$

where h = loss of head in bed of depth l with face velocity v and k = coefficient of permeability.

Rose (1945) used dimensional analysis to develop the equation

$$\frac{h}{l} = 1.067C_D \frac{v^2}{gd\psi} \frac{1}{f^4} \tag{13.2}$$

where f = bed porosity = volume of voids/total volume,
    d = characteristic diameter of bed particle,
    ψ = particle shape factor and
    C_D = Newton's drag coefficient = $(24/R) + (3/\sqrt{R}) + 0.34$

The Carman–Kozeny (Carman, 1937) equation produces similar results to those derived from Rose's equation

$$\frac{h}{l} = E \frac{(1-f)}{f^3} \frac{v^2}{gd\psi} \tag{13.3}$$

where $E = 150[(1-f)/R] + 1.75$.

The particle shape factor ψ in equations 13.2 and 13.3 is the ratio of the surface area of the equivalent volume sphere to the actual surface area of the particle, i.e.

$$\psi = \frac{A_0}{A} \tag{13.4}$$

where A_0 = surface area of sphere of volume V.

For spherical particles ψ is unity and the particle diameter $d = (6V/A)$. For other shapes $d = (6V/\psi A)$. Thus equations 13.2 and 13.3 can be rewritten as

$$\frac{h}{l} = 0.178C_D \frac{v^2}{gf^4} \frac{A}{V} \tag{13.5}$$

$$\frac{h}{l} = E \left(\frac{1-f}{f^3}\right) \frac{v^2}{g} \frac{A}{6V} \tag{13.6}$$

Typical values of ψ are given in Table 13.1.

Table 13.1 Typical values of particle shape factor

Material	ψ
Mica flakes	0.28
Crushed glass	0.65
Angular sand	0.73
Worn sand	0.89
Spherical sand	1.00

Worked example on filter head loss

A filter bed is made of 0.40 mm size angular sand and has an overall depth of 750 mm and a porosity of 42 per cent. Use the Rose formula to estimate the head loss of the clean bed at a filtration rate of 120 m/day. (Kinematic viscosity of water $= 1.01 \times 10^{-6}\,\mathrm{m^2/s}$.)

Filtration rate of 120 m/day $= 120/(60 \times 60 \times 24) = 1.39 \times 10^{-3}\,\mathrm{m/s}$

$R = 1.39 \times 10^{-3} \times 4 \times 10^{-4}/1.01 \times 10^{-6} = 0.55$, i.e. laminar flow

Hence C_D is given by equation 11.6

$$C_D = (24/0.55) + (3/0.55^{0.5}) + 0.34$$

$$= 48.01$$

Using equation 13.2

$$h/l = 1.067 \times 48.01 \times (1.39 \times 10^{-3})^2/(9.81 \times 4 \times 10^{-4} \times 0.73 \times 0.42^4)$$

$$= 1.110$$

i.e. head loss $= 1.110 \times 0.750 = 0.833\,\mathrm{m}$.

Filters are normally used with graded sand, e.g. 0.5–1.00 mm, so that it is necessary to obtain an average A/V value from

$$\left(\frac{A}{V}\right)_{av} = \frac{6}{\psi} \sum \frac{p}{d} \tag{13.7}$$

where p = proportion of particles of size d (from sieve analysis). The slow sand filter with low hydraulic loading (about $2\,\mathrm{m^3/m^2\,day}$) is cleaned by removal of the clogged surface layers and the bed is a homogeneous packing. In the case of the rapid filter (loading about $120\,\mathrm{m^3/m^2\,day}$) cleaning is by backwashing with filtrate from below, thus producing a stratified bed packing and it is necessary to take account of the variation of C_D with particle size. Thus for a rapid filter using Rose's equation

$$\frac{h}{l} = 1.067 \frac{v^2}{g\psi f^4} \sum C_D \frac{p}{d}$$

(13.8)

Examples of this method of calculation are set out in Rich (1961) and Fair *et al.* (1967).

13.2 Filter clogging

The equations above give the head loss with a clean bed, but when used for removal of suspended matter the porosity of the bed is continually changing due to the collection of particles in the voids. It is usually assumed that the rate of removal of particles is proportional to their concentration, i.e.

$$\frac{\partial c}{\partial l} = -\lambda c$$

(13.9)

where c = concentration of suspended solids entering the bed, l = depth from inlet surface and λ = a constant which is characteristic of the bed.

The equation is a partial differential because the solids concentration varies with time as well as position in the bed. Initially equation 13.9 may be integrated to give

$$\frac{c}{c_0} = e^{-\lambda_0 l}$$

(13.10)

where c_0 = solids concentration at bed surface ($l = 0$).

Work by Ives and Gregory (1967) has shown that if the bed retains solids the total head loss can be expressed as a linear function which for uni-size media is

$$H = h + \frac{Kvc_0 t}{(1 - f)}$$

(13.11)

where h = head loss, from the Carman–Kozeny equation, t = time of operation and K = a constant which relates to the particular bed.

For graded media the expression is similar, but with K replaced by another constant depending on the size grading and the variation of K with particle size.

Contrary to common belief the removal of suspended matter in a porous media bed is not simply a straining action. Removal of solids depends upon transport mechanisms and processes such as

- interception – where streamlines pass close enough to bed grains so that particles come into contact with bed grains
- diffusion – random Brownian movements can bring colloidal particles into the vicinity of a bed grain
- sedimentation – gravitational forces can move particles across streamlines into quiescent areas on upward-facing surfaces of bed grains, an analogy with the tray concept in Hazen's sedimentation theory
- hydrodynamic – particles in a velocity gradient often develop rotational movements which produce lateral forces capable of moving them across streamlines and providing flocculation.

Once transported into the pores of a bed the suspended matter is held there by attachment mechanisms due to physico-chemical and intermolecular forces similar to those which operate in coagulation. A bed of porous media is thus able to remove particles considerably smaller than the voids within the bed. A typical sand bed using 0.5–1.0 mm sand will have internal pores of around 0.1 mm but will trap particles as small as 0.001 mm, i.e. the size of bacteria.

Because of the complex nature of filtration processes it is not easy to predict the filtration behaviour of a suspension in purely mathematical terms. A measure of the filtration characteristics of a sample can be obtained using a small laboratory apparatus which indicates the increase in head loss across a bed of sand after successive applications of known volumes of the sample. The rate of increase of head loss with volume filtered is known as the 'filtrability index'.

In developing countries there has been some interest in development of a horizontal flow filter which utilizes relatively large media in the form of a series of gravel packs to provide pretreatment of turbid surfacewaters before conventional sand filtration. In such a system sedimentation plays a considerable role in the removal of suspended matter and a gravel bed with horizontal flow can be considered as having some similarity to the concepts of high rate settlement discussed in Chapter 11.

13.3 Filter washing

With a slow filter, penetration of solids is superficial and cleaning is achieved by removing the upper layer of the medium at intervals of a few months, washing and replacing when the bed depth falls below a specified value. The rapid filter clogs much more rapidly due to its higher hydraulic loading and the solids penetrate deep into the bed. Cleaning is achieved by backwashing at a rate of about ten times the normal filtration rate. The upward flow of water expands the bed producing a fluidized condition in which accumulated debris is scoured off the particles. Compressed air scouring prior to, or at the same time as, backwashing improves cleaning and reduces washwater consumption. Figure 13.2 illustrates the behaviour of a porous bed under backwash.

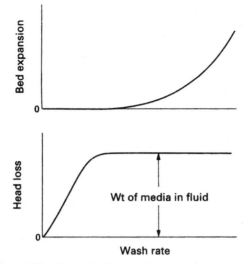

Figure 13.2   Behaviour of filter bed during backwashing.

As the backwash water is admitted to the bottom of the filter the bed begins to expand and there is an initial head loss. As the bed expands further the rate of increase of head loss decreases and when the whole bed is just suspended the head loss becomes constant. At this point the upward backwash force is equivalent to the downward gravitational force of the bed particles in water. Further increase in backwash flow increases the expansion but not the head loss. Excessive expansion is not desirable since the particles will be forced further apart, scouring action will be reduced and the backwash water consumption will be increased.

Referring to Figure 13.3 showing a bed under backwashing conditions, the expansion is $(l_e - l)/l$ (where l_e is expanded bed depth); this is usually 5–25 per cent in Europe although expansions up to 50 per cent are sometimes used in the USA. The washwater velocity is often termed the rise rate.

At the maximum frictional resistance by the bed, upward water force = expanded depth × net unit weight of medium × (volume of medium/total volume), i.e.

$$h g \rho_w = l_e (\rho_s - \rho_w) g (1 - f_e) \tag{13.12}$$

Therefore

$$\frac{h}{l_e} = \frac{(\rho_s - \rho_w)}{\rho_w} (1 - f_e) \tag{13.13}$$

or

$$\frac{h}{l_e} = (S_s - 1)(1 - f_e) \tag{13.14}$$

where S_s = specific gravity (relative density) of bed medium.

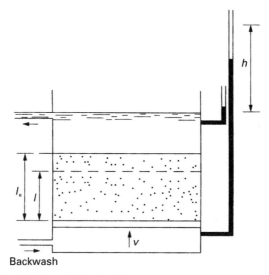

Backwash

Figure 13.3   Schematic filter under backwash.

The particles are kept in suspension because of the drag force exerted on them by the rising water. Thus from settling theory, equation 11.7, equating drag force and gravitational attraction, gives

$$C_D A \rho_w \frac{v^2}{s} \phi(f_e) = (\rho_s - \rho_w) g V \tag{13.15}$$

in which $\phi(f_e)$ is introduced because v is the face velocity of the backwash water whereas the drag is governed by the particle settling velocity v_s.

It has been found experimentally (Fair *et al.*, 1967) that

$$\phi(f_e) = \left(\frac{v_s}{v}\right)^2 = \left(\frac{1}{f_e}\right)^9 \tag{13.16}$$

Thus

$$f_e = \left(\frac{v}{v_s}\right)^{0.22} \tag{13.17}$$

or

$$v = v_s f_e^{4.5} \tag{13.18}$$

In practice the value of the exponent in equation 13.17 varies between about 0.2 and 0.4 depending upon the properties of the particular medium in use.

Now consider the static and fluidized conditions

$$(1 - f)l = (1 - f_e)l_e \tag{13.19}$$

$$\frac{l_e}{l} = \frac{1 - f}{1 - f_e} = \frac{1 - f}{1 - (v/v_s)^{0.22}} \tag{13.20}$$

For graded media it is necessary to use an arithmetic integration procedure to determine the overall expansion in the same way as when calculating head loss through a graded bed.

Because of the influence of viscosity on the drag forces in backwashing there is a significant temperature effect which is not always appreciated by designers and operators of rapid filters. For a given bed expansion the backwash rate at 4°C is only 75 per cent of that needed at 14°C and the backwash rate at 24°C needs to be 122 per cent of that at 14°C. If appropriate corrections are not made for this temperature effect cold climate operation may give excessive expansions with possible loss of media whereas in hot climates the expansion may not be sufficient for effective cleaning.

13.4 Types of filter

Two basic types of filter have been used in the water industry for many years and their main characteristics are summarized in Table 13.2. The slow sand filter (Figure 13.4) was the first type to be used and although some authorities consider the slow filter to be obsolete it does in fact still have many applications and may be particularly suitable in developing countries. Because of the low hydraulic loading there is only superficial penetration of suspended matter into the bed and runs of several weeks or months can be achieved with low turbidity inputs.

Due to the long run times in slow sand filters considerable biological activity occurs in a slime, the *schmutzdecke*, which forms on the surface of the bed. This slime layer contributes to the removal of fine suspended matter and often provides oxidation of organic contaminants in the raw water, which might otherwise cause taste and odour problems. This biological activity is seen as a positive benefit with raw waters which may contain pesticides and herbicides or other undesirable trace organics. A new development in slow filtration is to incorporate a layer of granular activated carbon (see Chapter 17) in the bed to increase removal of trace organics. Because of their large size the cleaning of

Table 13.2 Filter characteristics

Characteristic	Slow filter	Rapid filter
Filtration rate (m^3/m^2 day)	1–4	100–200
Bed depth (m)	0.8–1.0 (unstratified)	0.5–0.8 (stratified)
Effective size of sand	0.35–1.0	0.5–1.5
Uniformity coefficient	2.0–2.5	1.2–1.7
Max head loss (m)	1	1.5–2.0
Typical length of run	20–90 days	24–72 hours
Cleaning water (% of output)	0.1–0.3	1–5
Penetration of particles into bed	Superficial	Deep
Preceded by coagulation	No	Can be
Capital cost	High	Lower
Operating cost	Low	Higher
Power required	No	Yes

Note: Effective size is the 10 per cent by weight size; uniformity coefficient is ratio of 60 per cent by weight size to 10 per cent by weight size.

slow filters, which usually involves removing and washing the top few centimetres of the bed, is costly. It is therefore important that slow filters are not used for raw waters regularly having more than about 20 NTU turbidity. They are thus not suitable for use after chemical coagulation from which there is inevitably some carry-over of floc. With higher turbidity sources it is possible to utilize a double filtration process in which rapid filters remove most of the turbidity, the final removal and the oxidation of organic matter being achieved by secondary slow filters.

With suitable quality input water, either raw or pretreated, a slow filter should be able to produce a filtrate with

- less than 1 NTU turbidity
- 95 per cent removal of coliforms

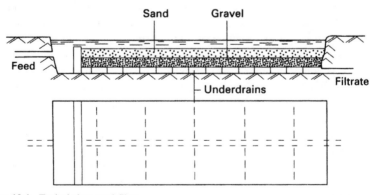

Figure 13.4 Typical slow sand filter.

- 99 per cent removal of *Cryptosporidium* and *Giardia* cysts
- 75 per cent removal of colour
- 10 per cent removal of TOC.

Rapid filters, which are widely used in water treatment and tertiary effluent treatment, normally operate under gravity conditions but may be installed as pressure filters in circumstances where, to preserve hydraulic head in the system, it is not desired to have a free water surface. Figure 13.5 shows the main features of a conventional rapid gravity filter (a pressure filter is essentially similar but enclosed in a steel vessel to withstand the operating head in the system).

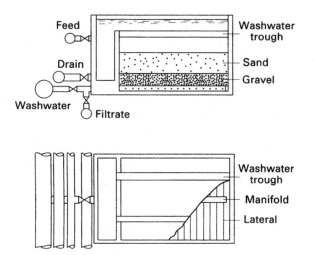

Figure 13.5   Conventional rapid gravity filter.

The normal method of filtration downward through a bed of stratified medium with the finest particles at the top is clearly inefficient since the main solids load falls on the smallest pores. A more logical method would be to have the larger particles at the top thus reserving the smaller voids to trap the really fine particles. This situation can be achieved by upward filtration in which the feed water passes up through the bed which is backwashed and stratified in the normal way so that the solids first meet the large bed particles. Care must be taken to control the filtration rate since high velocities will expand the bed and allow solids to escape. As turbidity breakthrough in an upflow filter tends to be sudden they are not popular for potable water treatment but have found application for tertiary filtration of wastewater effluents.

The use of a downflow filter comprising two media provides considerable benefits. A bed composed of a layer of anthracite (1.25–2.50 mm), which is less dense than the lower layer of 0.5 mm sand, will remain in this configuration after backwashing and again has the advantage of presenting the large bed particles to the feed first. Such a filter will operate at a much lower head loss than a sand bed of the same overall depth at the same hydraulic loading and without deterioration in filtrate quality. Some proprietary filters in the USA utilize multimedia beds with layers of light plastics, anthracite, sand and garnet to give improved utilization of void space although the additional cost and complexity of such beds is considerable.

When supplied with a raw or pretreated water with turbidity not more than 20 NTU at peak and 10 NTU on average a conventional rapid filter should achieve

- less than 1 NTU turbidity
- 90 per cent removal of coliforms
- 50–90 per cent removal of *Cryptosporidium* and *Giardia* cysts
- 10 per cent removal of colour
- 5 per cent removal of TOC

13.5 Filter operation and control

The head loss across a filter bed increases during the run and many beds incorporate a flow-control module which compensates for the increasing head loss in the bed so that the total head loss across the unit remains constant. Such control modules can sometimes malfunction and it is possible to operate filters without them, on a declining rate basis with a constant head or obtain a constant rate output by allowing the inlet head to increase as the run proceeds (Figure 13.6). These latter techniques are particularly appropriate for developing country installations where the complexity and expense of flow controllers may be

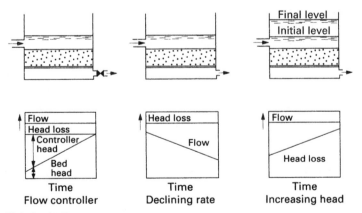

Figure 13.6 Rapid-filter control techniques.

undesirable. Whichever method of filter control is adopted it is important to ensure that the head loss across the bed does not increase to the point at which negative pressures will occur in the bed as shown in Figure 13.7. In such a situation the reduced pressure will allow air to come out of solution from the water and the bed may become blinded by air bubbles which impede the flow.

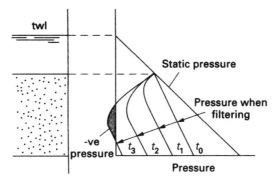

Figure 13.7 Build-up of head loss during filtration.

Filter runs may be terminated by one or more of the following criteria

- terminal head loss – a clean rapid filter bed will have an initial head loss of the order of 0.3 m and the run will probably need to be terminated when the head loss reaches about 2.5 m to prevent the occurrence of air blinding.
- filtrate quality – continuous monitoring of filtrate turbidity may be used to ensure that the run is stopped when the acceptable turbidity limit is exceeded, normally not greater than 1 NTU
- duration of run – a standard time interval of 24–72 h, determined by experience, may be satisfactory in conditions with more or less constant feed quality.

Whichever method is used, completely automatic operation of filters is possible and control systems are available which will take filters out of production and wash them in order of priority as determined by the operating criteria. Since the performance of a deep-bed filter is determined by several factors such as depth of bed, media size, filtration rate and water quality there are potential advantages to be gained from an optimized design procedure. The design can be based on experimental work with small diameter filters which provide the necessary data to enable the selection of physical parameters and operational criteria so that head loss and turbidity constraints are reached more or less simultaneously, as demonstrated in Figure 13.8.

Backwashing of rapid filters does of course use filtered water which would otherwise be available for supply and can account for around 2–3 per cent of

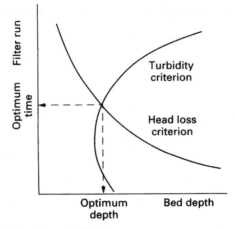

Figure 13.8 Optimization of deep-bed filtration.

output on average. Where water resources are limited it is common practice to treat backwash water by settlement to remove the suspended matter and return the supernatant water to the works inlet. This practice has caused some concern in circumstances where *Cryptosporidium* oocysts may be present in the raw water since they will be concentrated in the washwater, the return of which to the works may result in a high dose which is then not completely removed in the filters. Such potentially contaminated washwaters may be discharged to waste or alternatively passed through a membrane system to remove the oocysts before recycling to the plant inlet.

References

Carman, P. C. (1937). Fluid flow through granular beds. *Trans. Inst. Chem. Engnrs*, **15**, 150.

Fair, G. M., Geyer, J. C. and Okun, D. A. (1967). *Waste Water Engineering*: 2. *Water Purification and Waste-water Treatment and Disposal*, Ch. 27, pp. 1–49. New York: John Wiley.

Ives, K. J. and Gregory, J. (1967). Basic concepts of filtration. *Proc. Soc. Wat. Treat. Exam.*, **16**, 147.

Rich, L. G. (1961). *Unit Operations of Sanitary Engineering*, pp. 136–58. New York: John Wiley.

Rose, H. E. (1945). On the resistance coefficient–Reynolds number relationship for fluid flow through a bed of granular material. *Proc. Inst. Mech. Engnrs*, **153**, 145.

Further reading

Adin, A. and Rajagopalan, R. (1989). Breakthrough curves in granular media filtration. *J. Envir. Engng Am. Soc. Civ. Engnrs*, **115**, 785.

Adin, A., Baumann, R. E. and Cleasby, J. L. (1979). The application of filtration theory to pilot plant design. *J. Am. Wat. Wks Assn*, **71**, 17.

Amirtharajahm, A. (1988). Some theoretical and conceptual views of filtration. *J. Am. Wat. Wks Assn*, **80**(12), 36.

Bhargava, D. S. and Ojha, C. S. P. (1989). Theoretical analysis of backwash time in rapid sand filters. *Wat. Res*, **23**, 581.

Cleasby, J. L., Stangl, E. W. and Rice, G. A. (1975). Developments in backwashing of granular filters. *J. Envir. Engng Am. Soc. Civ. Engnrs*, **101**, 713.

Franklin, B. C., Hudson, J. A., Warnett, W. R. and Wilson, D. (1993). Three-stage treatment of Pennine waters: The first five years. *J. Instn Wat. Envir. Managt*, **7**, 222.

Glendenning, D. J. and Mitchell, J. (1996). Uprating water-treatment works supplying the Thames Water ring main. *J. C. Instn Wat. Envir. Managt*, **10**, 17.

Graham, N. J. D. (ed.) (1988). *Slow Sand Filtration*. Chichester: Ellis Horwood.

Ives, K. J. (ed.) (1975). *The Scientific Basis of Filtration*. Leyden: Noorhoff.

Ives, K. J. (1978). A new concept of filtrability. *Prog. Wat. Tech.*, **10**, 123.

Ives, K. J. (1989). Filtration studies with endoscopes. *Wat. Res.*, **23**, 861.

Rajapakse, J. P. and Ives, K. J. (1990). Pre-filtration of very highly turbid waters using pebble matrix filtration. *J. Instn Wat. Envir. Managt*, **4**, 140.

Saatci, A. M. (1990). Application of declining rate filtration theory—continuous operation. *J. Envir. Engng Am. Soc. Civ. Engnrs*, **116**, 87.

Tanner, S. A. and Ongerth, J. E. (1990). Evaluating the performance of slow sand filters in Northern Idaho. *J. Am. Wat. Wks Assn*, **82**(12), 51.

Tebbutt, T. H. Y. (1980). Filtration. In *Developments in Water Treatment*, Vol. 2 (W. M. Lewis, ed.). Barking: Applied Science.

Tebbutt, T. H. Y. and Shackleton, R. C. (1984). Temperature effects in filter backwashing. *Pub. Hlth Engnr*, **12**, 174.

Visscher, J. T. (1990). Slow sand filtration: Design, operation and maintenance. *J. Am. Wat. Wks Assn*, **82**(6), 67.

Wegelin, M. (1983). Roughing filters as pretreatment for slow sand filtration. *Water Supply*, **1**, 67.

Wegelin, M. (1996). *Surface Water Treatment by Roughing Filters: A Design, Construction and Operation Manual*. London: Intermediate Technology Publications.

Problems

1. A rapid gravity filter installation is to treat a flow of $0.5\,m^3/s$ at a filtration rate of $120\,m^3/m^2$ day with the proviso that the filtration rate with one filter washing is not to exceed $150\,m^3/m^2$ day. Determine the number of units and the area of each unit to satisfy these conditions. Each filter is washed for 5 min every 24 h at a wash rate of 10 mm/s, the filter being out of operation for a total of 30 min/day. Calculate the percentage of filter output used for washing. (5, $72\,m^3$, 2.6 per cent)

2. A laboratory-scale sand filter consists of a 10 mm diameter tube with a 900 mm deep bed of uniform 0.5 mm diameter spherical sand ($\psi = 1$), porosity 40 per cent. Determine the head loss using Rose's formula and the Carman–Kozeny formula when filtering at a rate of $140\,m^3/m^2$ day. For water at 20°C, $\mu = 1.01 \times 10^{-3}\,N\,s/m^2$. (676 mm, 515 mm)

3. The measured settling velocity of the sand particles in the filter in question 2 was 100 m/s. Determine the bed expansion when the filter is washed at a rate of 10 mm/s. (51 per cent)

14

Aerobic biological oxidation

The amount of organic matter which can be assimilated by a stream is limited by the availability of dissolved oxygen as discussed in Chapter 7. In industrialized areas where large volumes of wastewater are discharged to relatively small rivers, natural self-purification cannot maintain aerobic conditions and waste treatment additional to the removal of suspended matter by physical means is essential. Removal of soluble and colloidal organic matter can be achieved by the same reactions as occur in self-purification, but more efficient removal can be achieved in a treatment plant by providing optimum conditions.

14.1 Principles of biological oxidation

The fundamental speed of an aerobic oxidation reaction cannot readily be altered but by providing a large population of microorganisms in the form of a slime or sludge it is possible to achieve a rapid rate of removal of organic matter from solution. The large microbial surface permits initial adsorption of colloidal and soluble organics together with synthesis of new cells so that after a relatively short contact time the liquid phase contains little residual organic matter. The adsorbed organic matter is then oxidized to the normal aerobic end products (Figure 14.1).

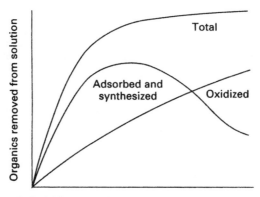

Figure 14.1   Removal of soluble organics in biological treatment.

The rate of removal of organic matter depends on the phase of the biological growth curve (Figure 6.2). Figure 14.2 shows the growth curve with an indication of the aeration times employed for various forms of aerobic treatment. BOD theory assumes a first-order reaction and although some reactions are thought to be first order there is evidence, sometimes conflicting, that other reactions are zero order (i.e. independent of concentration) or second order. The situation becomes more complex when dealing with wastes such as sewage which contain many different compounds.

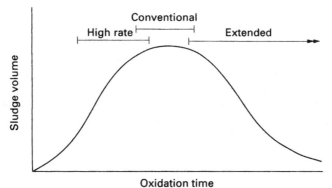

Figure 14.2 Operating zones for biological oxidation processes.

At high organic contents the reaction is likely to be of zero order with constant rate of removal of organics per unit cell weight, i.e.

$$\frac{Y}{S}\frac{dL}{dt} = K \tag{14.1}$$

where Y = mass of volatile suspended solids (VSS) synthesized/unit mass ultimate BOD removed (or per unit mass COD removed; see equation 6.6),
S = mass of VSS,
L = ultimate BOD and
K = constant.

When the organic concentration has been reduced to some limiting value the rate of removal becomes concentration dependent, i.e.

$$\frac{Y}{S}\frac{dL}{dt} = KL \text{ (first order)} \tag{14.2}$$

or

$$\frac{Y}{S}\frac{\mathrm{d}L}{\mathrm{d}t} = KL^2 \text{ (second order)} \tag{14.3}$$

About one-third of the COD of a waste is used for energy and the remaining two-thirds are utilized for synthesis of new cells. Thus sludge production measured as VSS ranges from 0.2 to 0.8 kg/kg COD removed depending on the substrate and time of aeration. Allowance must be made for any SS initially present in the waste.

The volatile solids accumulation is given by

$$\text{VSA} = (YL_r + cS_i - S_e)q - bS_m qt \tag{14.4}$$

where VSA = mass of VSS accumulated per unit time,
S_i = concentration of VSS in influent,
S_e = concentration of VSS in effluent,
S_m = concentration of VSS in system,
L_r = concentration of ultimate BOD removed per unit time,
c = fraction of non-biodegradable VSS in influent,
b = endogenous respiration constant per unit time,
q = rate of flow per unit time and
t = retention time of system.

For aerobic oxidation a typical value of Y would be 0.55 on an ultimate BOD basis (or $0.55 \times L/\text{BOD}$, for other than ultimate BOD values). A commonly used value of b is 0.15/day.

To maintain aerobic conditions in the reactor oxygen must be supplied, since it is utilized for oxidation reactions and for basic cell maintenance. The theoretical calculation of the oxygen requirement is complex and in most cases an empirical relationship is used. The oxygen requirement is thus a function of the COD removed and the endogenous respiration requirement and can be approximated by

$$O_2/\text{unit time} = 0.5L_r q + 1.42bS_m qt \tag{14.5}$$

The factor 0.5 allows for that portion of COD which is oxidized and the factor 1.42 is a typical conversion from VSS to COD for biological solids.

Worked example on biological oxidation

A sewage treatment plant has an average flow of $0.15 \, \text{m}^3/\text{s}$ and the incoming sewage has a COD of 750 mg/l and VSS of 400 mg/l. Primary sedimentation removes 40 per cent of the COD and 60 per cent of the SS. The settled flow is treated in an activated-sludge plant with a system VSS of 3000 mg/l and a retention time of 5 h. After final sedimentation the effluent is 50 mg/l COD and 20 mg/l VSS. Calculate the daily VSS accumulation assuming that 90 per cent of the incoming VSS are biodegradable, the synthesis constant (Y) is 0.55 and the endogenous respiration coefficient (b) is 0.12/day.

$$\text{Settled sewage contains } 750 \times 0.6 = 450 \text{ mg/l COD}$$

$$400 \times 0.4 = 160 \text{ mg/l VSS}$$

Now 1 mg/l = 1 g/m^3, so VSS accumulated/day, from equation 14.4, is

$$= [0.55(450 - 50) + 0.1(160) - 20] \times (0.15 \times 60 \times 60 \times 24) - 0.12(3000 \times 0.15 \times 60 \times 60 \times 5)$$

$$= 2\,799\,360 - 972\,000 = 1\,827\,360 \text{ g} = 1.827 \text{ tonnes}$$

14.2 Types of aerobic oxidation plant

Biological treatment reactors provide the high population of microorganisms in the form of either a fixed film on a suitable support surface or as a dispersed growth kept in suspension by an appropriate level of mixing.

There are four basic types of aerobic reactor

- biological filter, trickling filter or bacteria bed – fixed film systems
- activated sludge – dispersed growth systems
- oxidation pond – mainly dispersed growth systems
- land treatment – complex systems.

The biological filter and the activated-sludge process rely on similar principles, but the oxidation pond, which is more appropriate to warm sunny climates, utilizes symbiosis between algae and bacteria to produce the stabilization of organic matter together with a significant removal of faecal bacteria. Land treatment takes several forms and can be more of a 'natural' system rather than an engineered solution.

14.3 Biological filter

The oldest form of biological treatment unit consists basically of a bed of stone, circular or rectangular in plan (Figure 14.3) with intermittent or continuous addition of settled sewage to the surface. On a conventional filter the medium is 50–100 mm grading, preferably a hard angular stone, dosed by a rotating distributor mechanism, the normal depth of bed being 1.8 m.

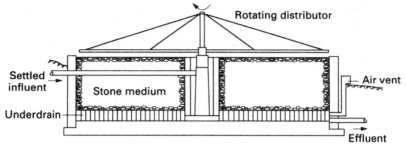

Figure 14.3 A conventional biological filter.

Liquid trickles through the interstices in the medium where microorganisms grow in the protected areas forming a slime or film, the liquid flowing over the film rather than through it (Figure 14.4). The microorganisms are attracted to the medium by van der Waals forces which are opposed by the shearing action of the liquid. Thus although there is little organic matter in solution in filter effluent there may be fairly high concentrations of SS in the form of displaced film. The filter effluent thus requires sedimentation in a humus tank to produce the desired effluent quality. The highest rate of oxidation takes place in the top section of the bed where the limiting factor is usually the amount of oxygen which can be supplied by natural ventilation (Figure 14.5).

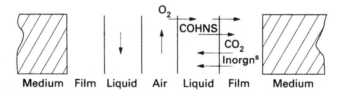

Figure 14.4 Idealized section of a biological filter.

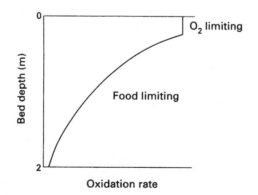

Figure 14.5 Relation between depth of filter and rate of oxidation.

Below this level the rate of oxidation decreases due to the reducing concentration of organic matter in the liquid phase, and there is normally little benefit in using depths of medium greater than 2 m. The liquid film may only be in contact with the microorganisms for a matter of 20–30 s, but because of the large surface area available this contact time is sufficient for adsorption and stabilization. The maximum rate of stabilization occurs at the microorganism/liquid interface since diffusion of organics through the film is slow. With a thick biological film, waste stabilization is not very efficient since much of the film is undergoing endogenous respiration.

Long experience in the UK has shown that to produce a 30 mg/l SS and 20 mg/l BOD effluent after humus settlement, filters treating domestic sewage should be loaded at $0.07–0.10 \, kg \, BOD/m^2$ day with a hydraulic loading of $0.12–0.6 \, m^3/m^3$ day. If the filter loading is increased the higher content of organic matter will promote heavy film growths which may result in blockage of the voids, causing ponding of the filter and anaerobic conditions. At conventional design loadings it is usual to obtain a fairly high degree of nitrification in the effluent in warm weather, although this will not be so apparent at the higher ranges of loading. Inevitably there have been many attempts during the years to produce more efficient biological filters which would operate at much higher loadings and several modifications of the basic process have been produced (Figure 14.6).

High-rate filtration

As mentioned previously, excessive loading on filters results in ponding; this can be obviated or reduced by using large filter media and hydraulic loadings of $1.8 \, m^3/m^3$ day will give 30:20 standard effluent from domestic sewage, albeit with little or no nitrification. If a less stabilized effluent is required, e.g. roughing treatment for strong industrial wastes, hydraulic loadings of up to $12 \, m^3/m^3$ day with organic loadings of up to $1.8 \, kg \, BOD/m^3$ day will give 60–70 per cent BOD removals. Treatment at such rates is facilitated by the use of plastic medium (90 per cent voids) in tall towers rather than the usual stone medium (40 per cent voids), the risk of ponding being thereby much reduced.

Alternating double filtration

A ponded filter can be brought back into use by applying the partially stabilized effluent from another filter. The film in alternating double filtration (ADF) alternately grows and disintegrates, the total amount of film being less than in a single filter so that higher rates of loading can safely be employed. Two filters are operated in series and when the first filter shows signs of ponding the order of flow through the filters is reversed. A second humus tank and additional pipework and pumping facilities are needed to operate ADF. A 30:20 standard effluent can be produced at loadings of $1.5 \, m^3/m^3$ day and

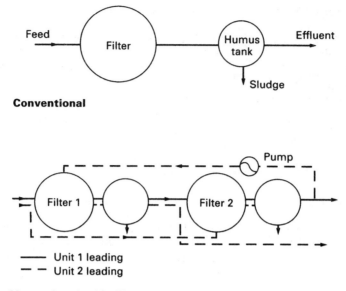

Conventional

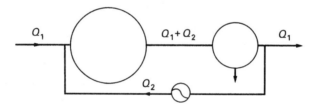

—— Unit 1 leading
— — Unit 2 leading

Alternating double filtration

Recirculation

Figure 14.6 Biological filtration process modifications.

0.24 kg BOD/m^3 day and the process is often useful in the relief of overloaded conventional biological filters.

Recirculation

This is another method for increasing filter capacity based on the principle of treating settled waste in admixture with settled filter effluent in ratios of 1:0.5–10 depending on the strength of the waste. The concentration of organic matter in the feed to the filter is thus reduced at the expense of larger hydraulic loadings and additional pumping and pipework. Recirculation may be used only at low flows, at a constant rate or at rates proportional to the incoming flow. With particularly strong wastes two-stage recirculation may be adopted.

Nitrifying filters

To provide further stabilization and nitrification for an activated-sludge plant, effluent high-rate filters are useful. At hydraulic loadings of up to $9m^3/m^3$ day and $0.05-0.1$ kg BOD/m^3 day a nitrified $30:20$ standard effluent can be produced. Nitrifying bacteria also find application in water treatment plants dealing with raw waters containing significant amounts of ammonia. The nitrifying organisms may be contained in a high-rate bacteria bed or in a vertical-flow floc-blanket settling tank. In either case there may be an added benefit in that the microorganisms can also oxidize trace organics present in the water which might otherwise cause taste and odour problems in the finished water.

Rotating biological contactors

The use of slowly rotating circular discs, often referred to as rotating biological contactors (RBCs) (Figure 14.7), provides a large surface area for film formation in a compact space. RBC units are usually factory-made tanks divided by baffles into a number of chambers in which the surfaces of the discs are regularly

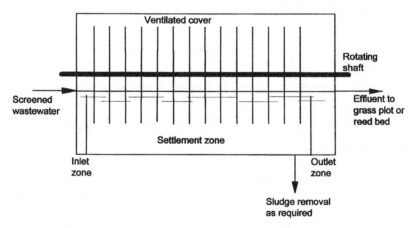

Figure 14.7   Rotating biological contactor.

immersed as they rotate. Discs are usually submerged to between 30 and 40 per cent of their depth. An inlet screen is essential and some installations incorporate a primary settling stage ahead of the disc zone. Heavy film growths soon occur under suitable loading conditions and high BOD removals can be achieved. Suspended matter in the incoming flow together with excess film settles in the bottom of the tank where some anaerobic stabilization will occur, although regular desludging is required to maintain proper operation of the system. It is usual for RBC units to be fitted with ventilated covers so that they are unobtrusive and they have become popular because of their reliable performance, even for quite small installations of a few houses. The speed of disc rotation should be in the range of 1–3 revs/min with a maximum peripheral velocity of 0.35 m/s. Typical organic loadings are up to 5 g BOD/m^2 day (based on the face area of the discs) but reduced to 2.5 g BOD/m^2 day if nitrification is required. The unit should have a minimum retention time of 1 h or, preferably, more for small populations where sudden variations in flow can occur.

Biological aerated filters

A relatively recent development is the biological aerated filter (BAF) (Figure 14.8) which exists in a number of forms somewhat similar to rapid gravity filters, with upflow, downflow or mixed flow regimes. The filter beds use plastics, expanded shale or sand media on which biological growth occurs and in which suspended solids are trapped. Backwashing is used to remove the retained solids as required.

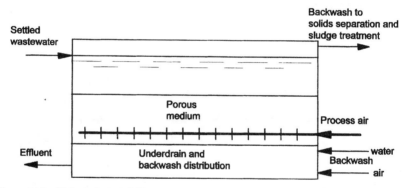

Figure 14.8 Biological aerated filter.

A final sedimentation stage is not needed since suspended solids remain in the bed until it is backwashed. Depending upon loading and operational conditions BAFs can produce nitrified or denitrified effluents. Typical organic loadings are up to 4 kg BOD/m^3 day to produce an effluent with around 100 mg/l COD. In tertiary nitrification mode, loadings of about 0.6 kg N/m^3 day can usually produce effluents with less than 0.5 mg/l ammonia nitrogen.

14.4 Activated sludge

This process depends on the use of a high concentration of microorganisms present as a floc kept suspended by agitation, originally with compressed air, although mechanical agitation is also now used (Figure 14.9). In either case high rates of oxygen transfer of around 2 kg O$_2$/kW h are possible. The effluent from

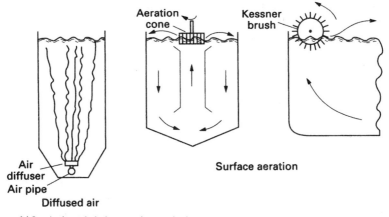

Figure 14.9 Activated sludge aeration methods.

the aeration stage is again low in dissolved organics but contains high SS (2000–8000 mg/l) which must be removed by sedimentation. The effectiveness of the process depends on the return of a portion of the separated sludge (living microorganisms) to the aeration zone to recommence stabilization. The initial attractions of the activated-sludge process were that it occupied much less space than a biological filter and had a much lower head loss. It has since proved useful for the treatment of many organic industrial wastes which were at one time thought to be toxic to biological systems.

In the diffused air system much of the air is used for agitation and only a small amount is actually utilized for the oxidation reactions. In the absence of agitation (as in the final settling tank) the solids quickly settle to the bottom and lose contact with the organic matter in the liquid stage; the settled solids rapidly become anaerobic if not returned to the aeration zone. Sufficient air must be transferred to the mixed liquor to maintain a DO of 1–2 mg/l. The mixed liquor must be of suitable concentration and activity to give rapid adsorption and oxidation of the waste as well as providing a rapidly settling sludge so that a clarified effluent is produced quickly and the sludge can be returned to the aeration zone without delay.

In general a sludge volume of 25–50 per cent of the flow through the plant is drawn off from the settling tank and between 50 and 90 per cent of this is returned to the aeration zone, the remainder being dewatered and disposed of along with other sludges from the plant. If insufficient sludge is returned, the mixed liquor suspended solids (MLSS) will be low and poor stabilization will result; the return of excessive amounts of sludge will result in very high MLSS which may not settle well and which may exert higher oxygen demands than can be satisfied. If sludge is not removed rapidly from the settling tanks rising sludge may occur due to the production of nitrogen by reduction of nitrates under anaerobic conditions – a very poor effluent is then produced.

Because of the importance of maintaining good-quality sludge in the process, various indices have been developed to assist in control. These are described below.

1. The sludge volume index, SVI, is given by

$$SVI = \frac{\text{settled volume of sludge in 30 min}(\%)}{\text{MLSS}(\%)} \qquad (14.6)$$

The SVI varies from about 40 to 100 for a good sludge, but may exceed 200 for a poor sludge with tendency to bulking. Bulking is used to describe a sludge with poor settling characteristics, often due to the presence of filamentous microorganisms which tend to occur in plants with an easily degradable wastewater low in nitrogen and where the DO in the mixed liquor is low. A problem which arises when using SVI as a measure of the settling properties of a mixed liquor is that the values obtained are affected by the solids concentration and the diameter of the vessel used for the test. To overcome these problems the SSV (stirred specific volume) test is carried out at a fixed fluid SS concentration of 3500 mg/l and in a standard 100 mm diameter vessel stirred at 1 rev/min.

2. The sludge density index, SDI, is given by

$$SDI = \frac{\text{MLSS}(\%) \times 100}{\text{settled volume of sludge in 30 min}(\%)} \qquad (14.7)$$

SDI varies from about 2 for a good sludge to about 0.3 for a poor sludge.

3. The mean cell residence time, θ_c, is given by

$$\theta_c \text{ (days)} = \frac{\text{aeration zone volume (m}^3\text{)} \times \text{MLVSS(mg/l)}}{\text{sludge wastage rate (m}^3\text{/day)} \times \text{sludge VSS (mg/l)}} \qquad (14.8)$$

For an activated-sludge plant producing a 30:20 standard effluent from normal settled sewage, conventional design criteria are 0.56 kg BOD/m³ day with a nominal retention time in the aeration zone of 4–8 h. Nominal air supply is about 6 m³/m³, retention in the final settling tank is usually about 2 h. Such a plant should reliably produce a 30:20 standard effluent although nitrification may not be complete. A degree of nitrogen removal can be achieved by mixing settled sewage and return sludge in an anaerobic tank ahead of the conventional aeration tank.

Many modifications of the activated-sludge process have been produced both in the form of aeration (fine and coarse bubble diffusers, high efficiency aerators with or without sparger rings) and in the actual process (Figure 14.10).

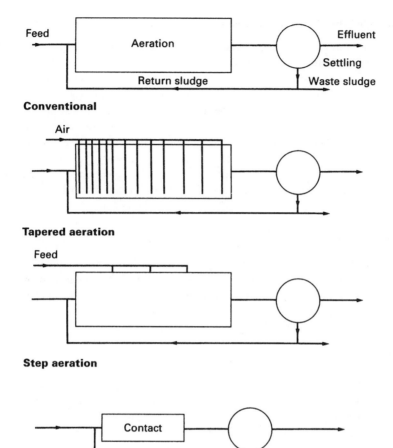

Figure 14.10   Modifications of the activated-sludge process.

High-rate activated sludge

With short retention times (2 h) and low MLSS (about 1000 mg/l), partial stabilization is achieved rapidly at low cost, at loadings of up to 1.6 kg BOD/ m^3 day with an air supply of about $3 m^3/m^3$. Such plants achieve BOD removals of 60–70 per cent and are suitable for pretreatment of strong wastes or for effluents discharged to estuarine waters where relaxed consent standards may be applied.

Tapered aeration and step aeration

In a conventional flow-through system the rate of oxidation is highest at the inlet end of the tank and it may sometimes be difficult to maintain aerobic conditions there if a uniform air distribution is used. With tapered aeration the air supply is progressively reduced along the length of the tank so that although the same total volume of air is used as before more of the air is concentrated at the tank inlet to cope with the high demand there. Step aeration aims to achieve the same object by adding the waste feed at intervals along the tank to give a more constant oxygen demand in the aeration zone. It is not then necessary to reaerate the return sludge before addition to the aeration zone which is something often necessary in conventional plants to prevent zero DO at the aeration basin inlet zone. The completely mixed unit is the natural extension of this concept.

Contact stabilization

This process uses the adsorptive capacity of the sludge to remove organic matter from solution in a small tank (30–60 min retention), the sludge and adsorbed organics then being transferred as a concentrated suspension to an aerobic digestion unit for stabilization (2–3 h retention). Solids in the contact zone are about 2000 mg/l, whereas in the digestion unit they may be as high as 20 000 mg/l.

Extended aeration

Using long aeration times (24–48 h) it is possible to operate in the endogenous respiration zone so that less sludge is produced than in a normal plant. A low organic loading is used, 0.24–0.32 kg BOD/m^3 day, and the plants have achieved some popularity for small communities where the reduced sludge volume and the relatively inoffensive nature of the mineralized sludge are considerable benefits. These benefits are, however, paid for in high operating costs (due to the long aeration time) and the plants do not normally produce a 30:20 standard effluent due to carry-over of solids from the settling zone.

Oxidation ditches

A development of the extended aeration process which has become very popular is the adoption of brush or paddle aerators to provide aeration and to create motion in continuous ditches which can be relatively cheap to construct in suitable ground conditions. Settlement may be achieved by intermittent shut down of the aerator although on larger units a separate continuous-flow settling tank is usually provided.

Sequencing batch reactors (SBR)

These are simple installations which utilize a fill-and-draw mode of operation rather than the normal continuous flow mode. In this they are a reversion to the original experiments on the activated sludge process by Arden and Lockett. SBRs have become popular in Australasia and for small plants in several other parts of the world although they are not as yet widely used in the UK. The process involves at least two identical tanks to which the incoming flow is directed alternately. When a tank is full it is aerated for a period of around 10–12 h, the air is then turned off to allow sedimentation to occur after which the supernatant effluent is decanted. During the aeration and settling periods the incoming flow is switched to the other tank to allow it to fill up after which the aeration and sedimentation stages occur. Because batch settlement usually produces a lower SS concentration than continuous flow settlement a correctly operated SBR can often produce a high quality effluent of less than 10 mg/l BOD and SS and 2 mg/l ammonia nitrogen from domestic sewage. Process control is normally by simple timers although a sensitive sludge level detector is helpful in preventing loss of settled solids in the effluent. The build-up of surplus activated sludge is controlled by regular removal of excess sludge as required.

Pure oxygen activated sludge

A recent development has been the introduction of activated-sludge plants operated with pure oxygen. Such installations involve the introduction of oxygen into closed stirred reaction tanks. It is claimed that these units can operate at relatively high MLSS levels (6000–8000 mg/l) whilst providing good sludge-settling characteristics and giving economies in power consumption and land area requirements. Operational data from pilot plants in the UK have not always confirmed the original claims, however, and they are probably only economic for very strong organic wastewaters with high oxygen demands.

Fluidized beds

In attempts to increase the population of microorganisms in a biological reactor, hence increasing the efficiency of the process, a number of systems have been developed in which a bed of granular material such as sand or plastics is supported in a tubular reactor by a controlled upward flow. The upward flow is made up of incoming wastewater plus a variable effluent recycle. Micro-organisms in the flow colonize the particles of medium which provide a very large surface area for film growth.

Deep shaft

This process, which was originally developed for single cell protein production, has been utilized in a number of installations for wastewater treatment. The process depends upon the hydrostatic pressure generated at the bottom of a shaft 100–150 m deep to enhance oxygen transfer to a rapidly circulating dispersed growth liquor. Hydraulic retention times are usually 1–2 h with a sludge age of around four days and BOD removal reaches 90 per cent.

Thermophilic aerobic digestion (TAD)

This is a relatively new process which has received some attention as an alternative method of stabilizing organic sludges with high organic contents. Conventional sludge stabilization uses anaerobic digestion as described in Chapter 15 but TAD appears to offer more economic treatment for small plants up to 10 000 population equivalent. The process requires an operating temperature of around 55°C which can be self-sustained if the reactor vessel is well insulated and the sludge has a volatile solids content of more than 3 per cent. The amount of oxygen required for the process is governed by the COD removal required to produce stabilization of the sludge. COD removals of around 50 per cent can be achieved with retention times of 10–15 days but the air supply has to be closely matched to the demand for reliable operation.

Sludge production

It is important to appreciate that because biological treatment systems involve both oxidation and synthesis there will always be some biomass which is discharged from the biological stage. Thus although the effluent from a biological reactor should be low in dissolved BOD it will contain significant amounts of suspended matter with associated BOD. Most aerobic biological processes utilize a secondary sedimentation stage to remove the biological solids and, in the case of activated sludge systems, recover them for recycling. It is thus sensible to consider the combination of the biological stage and the solids/liquid separation stage as being integral parts of the overall process. Many operational problems with activated sludge plants are due to poor solids settling characteristics which make clarification difficult. Some work has been carried out using reactors where the solids/liquid separation is achieved by passage through membranes which retain the suspended solids and allow the liquid effluent to pass out of the system. Such membrane reactors may be able to operate at high MLVSS levels, but they are still at the development stage.

14.5 Oxidation pond

Oxidation or waste stabilization ponds are shallow constructions, usually receiving raw sewage, which provide treatment by natural stabilization processes in suitable climatic conditions. Given sufficient land area they can give a very satisfactory form of wastewater treatment in warm sunny climates. Although usually considered to be best suited to hot climates it is possible to use oxidation ponds successfully in temperate regions. They are cheap to construct, simple to operate and provide good removals of organic matter and pathogenic micro-organisms but can release high concentrations of algae in the effluent with detrimental effects on the BOD and SS levels. A significant disadvantage of oxidation ponds in areas where land costs are high is the need to provide around $10\,m^2$ of pond area per person served. In some situations ponds may operate without producing an effluent due to evaporation and seepage, but in most cases they are designed as continuous flow systems.

Four main types of pond are used

- facultative ponds
- maturation ponds
- anaerobic ponds
- aerated ponds.

Facultative ponds are by far the most common and, as the name implies, combine aerobic and anaerobic activity in the same unit. Chlorophyll-bearing microorganisms, phytoflagellates and algae operate in these ponds by utilizing the inorganic salts and carbon dioxide provided by the bacterial decomposition of organic matter as shown in Figure 14.11. The oxygen produced by photo-

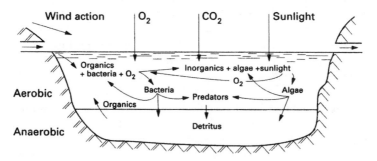

Figure 14.11   Reactions in a facultative oxidation pond.

synthesis, which may give DO levels of $15-30\,mg/l$ in late afternoon, is available for aerobic bacteriological activity although the DO level will fall during the night and may reach zero if the pond is overloaded. In the bottom deposits, anaerobic activity produces some stabilization of the sludge and releases some of the organic matter in soluble form for further degradation in the aerobic zone. Facultative ponds are usually $1-2\,m$ deep with a surface loading of $0.02-0.05\,kg\,BOD/m^2$ day and nominal retention times of $5-30\,day$, although these values should be modified for extreme temperatures. Because of the relatively long retention times and low organic concentration in such ponds there is a considerable removal of bacteria by endogenous respiration and by settlement. Bacteria and phytoplankton are preyed upon by ciliates, rotifers and crustaceans, but some will escape in the effluent. Heavy algal growths occur and their presence in the effluent will produce moderate to high SS levels unless some means of harvesting or removal is employed. BOD removals of 70–85 per cent are possible, although algae in the effluent can significantly increase the BOD and SS levels. The most satisfactory shape for ponds is rectangular with a length:breadth ratio of about 3:1. Simple earth banks are usually sufficient although in large ponds, wave action may make bank protection with paving slabs or similar materials desirable. Shallow areas at the edges should be avoided to discourage mosquitoes and grass/weed cutting together with occasional insecticide spraying may be necessary. Solids will accumulate in the pond at a rate of 0.1–0.3 m³/person year so that desludging will only be required at relatively long intervals of several years. Facultative ponds are widely used and their performance has been studied in some detail so that a number of design relationships are available. One of the most popular is that developed by McGarry and Pescod as an empirical fit to a wide range of performance data

$$\text{allowable kg BOD/ha day} = 60.3 \times 1.0993^T \qquad (14.9)$$

where T = minimum mean monthly air temperature (°C).

Maturation ponds are shallow, fully aerobic ponds with a very low organic loading (<0.01 kg BOD/m² day) used primarily as a secondary stage of treatment following a facultative pond or other biological treatment unit. Again, large algal growths occur but their most important feature is the high removal of pathogenic bacteria because of the unfavourable environment for such organisms in the pond.

Anaerobic ponds are operated with a fairly high organic loading of about 0.5 kg BOD/m² day with a depth of 3–5 m to ensure anaerobic conditions. They are capable of giving 50–60 per cent BOD removal with a retention time of around 30 days and may be suitable for pretreating strong organic wastes before addition to facultative ponds. Anaerobic ponds are likely to produce odours so that they should not be sited near populated areas.

Aerated ponds are analogous to the activated-sludge extended aeration process and utilize floating aerators to maintain DO levels and to provide mixing. BOD loadings of around 0.2 kg/m² day are possible with retention times of a few days and good-quality effluents can be produced. The process involves the maintenance of an essentially bacterial floc rather than the bacterial/algal system of the simpler ponds. The need for mechanical plant and reliable power supply detracts from the basic simplicity of the oxidation pond but aerated ponds can have applications in the urban areas of developing countries. Some aerated ponds have been operated with wind-powered aerators which, in suitable environments, can provide effective mixing and oxygen transfer.

14.6 Land treatment

The concept of land treatment of wastewater has a long history stretching as far back as the original 'sewage farms' in which crude sewage was discharged to ploughed land. As long as application rates were kept low and the soil was reasonably permeable the practice was fairly effective, although it did pose potential health hazards and was a possible source of groundwater pollution. Overland treatment and infiltration treatment became unpopular in the early part of this century largely as a result of the undesirable consequences of excessive loading of the natural soil systems as populations increased. There has, however, been some resurgence of interest in land treatment of sewage during the last few years. Although there are some simple overland flow and infiltration systems, operated at appropriate loadings most of the developments have been in relation to reed bed systems. These are sometimes referred to as root zone systems. The concept involves the construction of an engineered bed of various wetland plants with a predominance of reeds of the *Phragmites* species. The plants and their root systems provide appropriate hydraulic flow characteristics through the bed and help to transfer oxygen from the atmosphere into the soil or gravel in the bed (Figure 14.12).

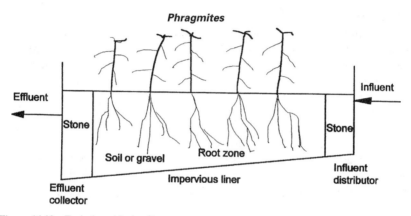

Figure 14.12 Typical reed bed unit.

Aerobic oxidation of the organic matter in the wastewater flowing through the bed thus takes place although the presence of anaerobic pockets may also allow some denitrification to occur. There is also the possibility of some phosphate removal by chemical reactions in the bed. When designed on the basis of $5\,m^2$ per head, depending upon the particular conditions, it is claimed that a reed bed unit can give very high quality effluents with low levels of nutrients and high removals of faecal bacteria. Reed bed systems have been constructed in Germany and several other countries in Europe although initial trials in the UK using them to provide full treatment did not always produce the expected effluent quality. However, reed beds have become widely used by Severn Trent Water and other UK water companies to provide tertiary and stormwater treatment for rural works, often incorporating RBC units. In these circumstances a reed bed area of 0.5 to $1\,m^2$/person is appropriate and they are able to ensure production of a nitrified effluent under most operating conditions.

Further reading

Arden, E. and Lockett, W. T. (1914). Experiments on the oxidation of sewage without the aid of filters. *J. Soc. Chem. Ind.*, **33**, 10.

Baker, J. M. and Graves, Q. B. (1968). Recent approaches for trickling filter design. *J. San. Engng Div. Am. Soc. Civ. Engnrs*, **94**, 61.

Barnes, D., Forster, C. F. and Johnstone, D. W. M. (eds) (1982). *Oxidation Ditches in Wastewater Treatment*. London: Pitman.

Bayes, C. D., Bache, D. H. and Dickson, R. A. (1989). Land-treatment systems: Design and performance with special reference to reed beds. *J. Inst. Wat. Envir. Managt*, **3**, 588.

Butler, J. E., Loveridge, R. F., Ford, M. G. *et al.* (1990). Gravel bed hydroponic systems used for secondary and tertiary treatment of sewage effluent. *J. Instn Wat. Envir. Managt*, **3**, 276.

Christoulas, D. G. and Tebbutt, T. H. Y. (1982). A simple model of the complete-mix activated sludge process. *Envr. Tech. Ltrs*, **3**, 89.

Cooper, P. F. and Atkinson, B. (1981). *Biological Fluidised Bed Treatment Of Water and Wastewater*. Chichester: Ellis Horwood.

Cooper, P. F. and Wheeldon, D. H. V. (1980). Fluidized and expanded-bed reactors for waste-water treatment. *Wat. Pollut. Control*, **79**, 286.

Cooper, P. F., Hobson, J. A. and Jones, S. (1989). Sewage treatment by reed bed systems. *J. Instn Wat. Envir. Managt*, **3**, 60.

Curi, K. and Eckenfelder, W. W. (eds) (1980). *Theory and Practice of Biological Wastewater Treatment*. Alpen aan den Fijn: Sitjhoff and Hordhoff.

Edgington, R. and Clay, S. (1993). Evaluation and development of a thermophilic aerobic digester at Castle Donington. *J. Instn Wat. Envir. Managt*, **7**, 149.

Ellis, K. V. and Banaga, S. E. I. (1976). A study of rotating-disc treatment units operating at different temperatures. *Wat. Pollut. Control*, **75**, 73.

Horan, N. J. (1990). *Biological Wastewater Treatment Systems*. Chichester: John Wiley.

Irwin, R. A., Brignal, W. J. and Biss, M. A. (1989). Experiences with the deep-shaft process at Tilbury. *J. Instn Wat. Envir. Managt*, **3**, 280.

Jones, G. A. (1991). Comparison of alternative operating modes on the Halifax activated-sludge plant. *J. Wat. Envir. Managt*, **5**, 43.

Lilly, W., Bourn, G., Crabtree, H. *et al.* (1991). The production of high quality effluents in sewage treatment using the Biocarbone process. *J. Instn Wat. Envir. Managt*, **5**, 123.

Mara, D. D. and Pearson, H. W. (1987). *Waste Stabilization Ponds: Design Manual for Mediterranean Europe*. Copenhagen: World Health Organization.

Mara, D. D., Mills, S. W., Pearson, H. W. and Alabaster, G. P. (1992). Waste stabilization ponds: a viable alternative for small community treatment systems. *J. Instn Wat. Envir. Managt*, **6**, 72.

Marecos do Monte, M. H. (1992). Waste stabilization ponds in Europe. *J. Instn Wat. Envir. Managt*, **6**, 73.

McGarry, M. G. and Pescod, M. B. (1970). Stabilization pond design criteria for tropical Asia. *Proc. Second International Symposium on Waste Treatment Lagoons*, Kansas City.

Meaney, B. (1994). Operation of submerged filters by Anglian Water Services Ltd. *J. Instn Wat. Envir. Managt*, **8**, 327.

Padukone, N. and Andrews, G. F. (1989). A simple, conceptual mathematical model for the activated sludge process and its variants. *Wat. Res.*, **23**, 1535.

Pullin, B. P. and Hammer, D. A. (1991). Aquatic plants improve wastewater treatment. *Wat. Envir. Tech.*, **3**(3), 36.

United States Environmental Protection Agency (1988). *Constructed Wetlands and Aquatic Plant Systems for Municipal Wastewater Treatment*. Cincinnati: USEPA.

Upton, J. E., Green, M. B. and Findlay, G. E. (1995). Sewage treatment for small communities: the Severn Trent approach. *J. Instn Wat. Envir. Managt*, **9**, 64.

Ware, A. J. and Pescod, M. B. (1989). Full-scale studies with an anaerobic/aerobic RBC unit treating brewery wastewater. *Wat. Sci. Tech.*, **21**, 197.

White, M. J. D. (1976). Design and control of secondary settlement tanks. *Wat. Pollut. Control*, **75**, 419.

Wood, L. B., King, R. P., Durkin, M. K. *et al.* (1976). The operation of a simple activated sludge plant in an atmosphere of pure oxygen. *Pub. Hlth Engnr*, **4**, 36.

World Health Organization (1987). *Wastewater Stabilization Ponds: Principles of Planning and Practice*. Alexandria: WHO.

Problems

1. An activated-sludge plant with 3000 mg/l MLVSS treats a waste with an ultimate BOD of 1000 mg/l having 350 mg/l VSS which are 90 per cent biodegradable. The plant effluent contains 30 mg/l ultimate BOD and 20 mg/l VSS. The hydraulic retention time of the system is 6 h. Determine the daily VSS accumulation and the oxygen requirement for a flow of $0.1\,m^3/s$ if the synthesis constant (Y) is 0.55 and the endogenous respiration constant (b) is 0.15/day. (3767 kg/day, 5570 kg/day)

2. Control analyses on an activated-sludge plant indicated MLSS 4500 mg/l and 25 per cent solids settled in 30 min. The plant treats a waste with 300 mg/l SS and a flow of $0.1\,m^3/s$. Aeration zone capacity is $2500\,m^3$. Sludge wastage rate $100\,m^3/day$ with VSS 15 000 mg/l. Calculate the SVI, SDI, and mean cell residence time. (55.5, 1.8, 7.5 days)

3. Compare the area requirements for conventional filters $(0.1\,kg\,BOD/m^3\,day)$ and conventional activated sludge $(0.56\,kg\,BOD/m^3\,day)$ for the flow from a town of 28 000 population, d.w.f. 200 l/person day with 250 mg/l BOD. Assume a filter depth of 2 m and 3 m deep aeration tanks. Primary sedimentation removes 35 per cent of the applied BOD. $(4550\,m^2, 325\,m^2)$

4. A West African village has a population of 500 people, a daily wastewater flow of 45 l/person and an individual BOD contribution of 0.045 kg/day. Determine the surface area required for facultative oxidation pond treatment of the village's wastewater assuming that the minimum mean monthly air temperature at the site is 10°C. (0.145 ha)

15

Anaerobic biological oxidation

With very strong organic wastes containing high suspended solids and with the sludges from primary sedimentation and biological treatment it becomes difficult to maintain aerobic conditions. The physical limitations of oxygen transfer equipment may prevent satisfaction of the oxygen demand with consequent onset of anaerobic conditions. In such circumstances it may be more appropriate to achieve partial stabilization by anaerobic oxidation or digestion.

15.1 Principles of anaerobic oxidation

Anaerobic oxidation obeys the same general laws as aerobic oxidation so that equations 14.1, 14.2, 14.3 and 14.4 may be applied. The methane produced by anaerobic oxidation is of some value as a fuel and the volume produced from a particular organic compound can be determined from the following relation

$$C_nH_aO_b + \left(n - \frac{a}{4} - \frac{b}{2}\right) H_2O \rightarrow \left(\frac{n}{2} - \frac{a}{8} + \frac{b}{4}\right) CO_2 +$$

$$\left(\frac{n}{2} + \frac{a}{8} - \frac{b}{4}\right) CH_4 \qquad (15.1)$$

At standard temperature and pressure (STP), 1 kg ultimate BOD (or 1 kg COD) oxidized anaerobically yields about 0.35 m³ methane gas which has a calorific value of 35 kJ/l. In a well operated anaerobic process the methane content of the gas is usually around 65 per cent with carbon dioxide making up the remainder except for traces of hydrogen sulphide, nitrogen and some water vapour.

The rate of gas production is temperature dependent as shown in Figure 15.1. Optimum gas production occurs at 35°C (mesophilic digestion) and 55°C (thermophilic digestion). The higher temperature is normally only economic in warm climates because of the high heat loss in other regions.

196

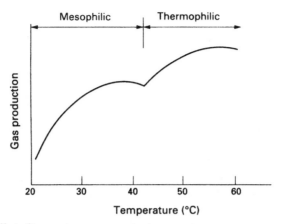

Figure 15.1   Effect of temperature on gas production.

In practice when allowing for synthesis, methane production can be estimated from the formula

$$G = 0.35(L_r q - 1.42 VSA) \tag{15.2}$$

where G = volume of CH_4 produced (m^3 per unit time),
  L_r = concentration of ultimate BOD removed per unit time,
  q = rate of flow per unit time and
  VSA = mass of VSS accumulated per unit time.

The mass of VSS produced per unit time in an anaerobic process can be obtained from

$$VSA = \frac{YL_r q}{1 + bt_s} \tag{15.3}$$

where Y = mass of VSS synthesized per unit mass ultimate BOD removed, b = endogenous respiration constant per unit time and t_s = solids retention time = mass of solids in system/solids accumulation per unit time.

Worked example on anaerobic digestion

An anaerobic digester is to be used to stabilize the effluent from an organic chemical plant. The discharge has a flow of $250\,m^3$/day with an ultimate BOD of $9000\,mg/l$. The digester is expected to remove 85 per cent of the incoming BOD and the values of the synthesis constant (Y) and the endogenous respiration constant (b) are 0.09 and 0.011/day respectively. Calculate the daily volatile solids accumulation if the solids retention time is 100 days.

Using equation 15.3,

$$VSA/day = 0.09 \times 0.9 \times 9000 \times 250/[1 + (0.011 \times 100)]$$
$$= 18\,250/2.1\,g$$
$$= 86.78\,kg$$

It will be noted that equation 15.3 includes terms which depend upon the solids retention time on both sides of the equation since the solids retention time is related to the solids accumulation. In the absence of information about the solids retention time it is necessary to establish the solids present in the digester and then solve for solids accumulation by trial and error.

15.2 Applications of anaerobic treatment

Sludge digestion

The main conventional use of anaerobic treatment is for digestion of the primary and secondary sludges produced from the treatment of domestic wastewater. These sludges have solids contents of between 20 000 and 60 000 mg/l (2–6 per cent) about 70 per cent of which are organic in origin. Primary sludges, which usually contain co-settled secondary biological sludges, are readily putrescible with a heterogeneous appearance and a heavy faecal odour. The effect of anaerobic stabilization is to reduce the volatile content to less than 50 per cent and the total solids to about two-thirds of the original value. The digested sludge is homogeneous in character with relatively stable characteristics and a tarry odour. Because of the homogenization of the sludge during the mixing processes in the reactor it is not uncommon to find that digested sludges are difficult to dewater. They are useful as soil conditioners since they contain considerable amounts of nutrients and the anaerobic reaction with the associated elevated temperatures is reasonably effective at removing pathogenic microorganisms from the sludge.

Conventional anaerobic digestion is carried out as a two-stage process (Figure 15.2), with the first, contact, stage heated to the required temperature by burning some of the methane gas produced in the process. After a suitable residence time most of the gas will have been evolved and the digested sludge is transferred to an unheated secondary stage where solids/liquid separation takes place. The supernatant liquor, which is high in soluble organics (up to 10 000 mg/l BOD), is drawn off for recycling to the inlet of the main treatment plant where it may constitute up to l0 per cent of the incoming BOD load. The consolidated sludge is removed for further treatment and disposal. Common design criteria for sludge digesters are volatile solids loadings of 0.5–$1.0\,kg/m^3$ day with volumetric retention times of 10–20 days. Effective mixing of the contents of the primary

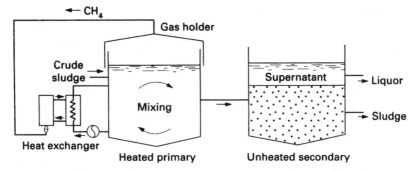

Figure 15.2 Conventional sludge-digestion plant.

digester by mechanical stirrers or by gas recirculation is essential for good performance. Most sludge digesters operate under mesophilic conditions but in warm climates thermophilic operation may be feasible.

Industrial waste treatment

With some strong organic wastewaters, notably those from some food processing operations and from the manufacture of alcoholic drinks, the high oxygen demand makes it difficult to achieve aerobic conditions in a treatment plant. In such circumstances anaerobic treatment can provide an attractive means of removing much of the BOD load in a relatively small installation. The energy value of the methane produced by the anaerobic reactions can defray some of the costs of the waste treatment operation. Because of its nature the anaerobic process cannot remove the BOD down to low levels, as can be achieved by aerobic reactions, so that the effluent from an anaerobic plant will be unlikely to be suitable for direct discharge to a watercourse. The effluent can, however, be discharged to municipal sewers at a much lower cost or alternatively passed to a small aerobic treatment stage before discharge to a receiving water. If discharged to a sewer the consent may require inhibition of the methane-forming bacteria because continued methane production could result in an explosion hazard in the sewer.

Because of economic factors, which are of particular importance for industrial wastewater treatment, a number of variants of anaerobic process systems have been developed in recent years. In addition to the conventional stirred tank used in most sludge digesters it is now possible to use a number of high-rate systems with loading rates of up to $20 \, kg \, VS/m^3$ day and which can give BOD removals of up to 90 per cent with suitable wastewaters. These high-rate systems may be in the form of

- continuously stirred contact digesters with more effective solids/liquid separation units
- submerged anaerobic filters (upflow or downflow modes)

- fluidized beds in which biomass is grown on an inert material such as sand or lighter materials such as pumice or PVC
- upflow anaerobic sludge blanket (UASB) systems where a stable blanket of granular sludge is formed.

In Europe there are around 400 industrial installations using anaerobic treatment, with the UASB process being the most common although initial formation of granular solids has often proved difficult unless sludge from an actively granulating plant is brought in as a seed material.

It should be appreciated that with these, and most other high-rate systems, it is likely that the process will be more difficult to operate and will be more sensitive to inhibitory constituents in the wastewater than more lightly loaded systems. Effective process control is thus essential and the composition of the influent must be regularly monitored to reduce the risk of process failures.

15.3 Operation of digesters

As outlined in Chapter 6 the anaerobic process is sensitive to acid pH conditions and requires careful control.

For good digestion the pH is usually between 6.5 and 7.5 and a falling pH means that the process is becoming unbalanced. Excess production of volatile acids destroys the buffering capacity of alkalinity in the sludge, lowers the pH and reduces gas production and its methane content (Figure 15.3). As long as the

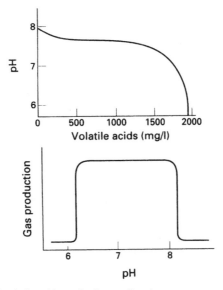

Figure 15.3 Effect of volatile acids production on digestion.

sludge has a fairly high alkalinity an increase in acid production may initially produce little effect on pH so that measurement of volatile acids is a better control parameter. The normal volatile acid content is 250–1000 mg/l and if it exceeds 2000 mg/l, trouble is likely. Lime is often used to aid recovery of digestion after high acid production but it is better to control the organic load and pH to prevent overproduction of volatile acids.

The changes which occur in a simple batch-fed digester are shown in Figure 15.4. The initial drop in pH occurs because of the faster action of the acid-forming bacteria. As the methane formers build up, the acid content is reduced and gas production increases as does the methane content of the gas. The initial start-up of digestion is usually achieved by seeding with active sludge from another plant, or by starting with partial load (about one-tenth of the normal) and slowly increasing. These methods should prevent excessive volatile acid production which could inhibit the growth of methane bacteria.

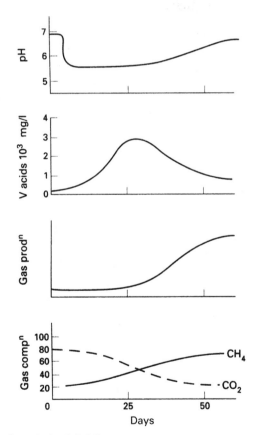

Figure 15.4 Behaviour of a batch-fed digester.

In small treatment plants digestion is sometimes carried out in unheated tanks with no facilities for gas collection. Such a procedure is only satisfactory in warm climates since in temperate countries active digestion occurs only during the summer. The septic tank used for single houses and small communities is in fact an anaerobic oxidation plant which removes suspended solids from sewage and breaks them down anaerobically. Septic tank effluent, whilst low in SS, still has a high BOD and should be treated on a biological filter before discharge to a watercourse. There is a build-up of solids in a septic tank which normally means that desludging is required about twice a year.

Further reading

Anderson, G. K. and Donnelly, T. (1977). Anaerobic digestion of high strength industrial wastewaters. *Pub. Hlth Engnr*, **5**, 64.

Andrew, P. R. and Salt, A. (1987). The Bury sludge digestion plant: Early operating experiences. *J. Instn Wat. Envir. Managt*, **1**, 22.

Brade, C. E. and Noone, G. P. (1981). Anaerobic sludge digestion—need it be expensive? Making more of existing resources. *Wat. Pollut. Control*, **80**, 70.

Bruce, A. M. (1981). New approaches to anaerobic sludge digestion. *J. Instn Wat. Engnrs Scientists*, **35**, 215.

Mosey, F. E. (1982). New developments in the anaerobic treatment of industrial wastes. *Wat. Pollut. Control*, **81**, 540.

Noone, G. P. and Brade, C. E. (1982). Low-cost provision of anaerobic digestion: II. High-rate and prefabricated systems. *Wat. Pollut. Control*, **81**, 479.

Noone, G. P. and Brade, C. E. (1985). Anaerobic sludge digestion—need it be expensive? III. Integrated and low-cost digestion. *Wat. Pollut. Control*, **84**, 309.

Sanz, I. and Fernandez-Planco, F. (1989). Anaerobic treatment of municipal sewage in UASB and AFBR reactors. *Envir. Technol. Lett.*, **10**, 453.

Speece, R. E. (1988). A survey of municipal anaerobic sludge digesters and diagnostic activity surveys. *Wat. Res.* **12**, 365.

Wheatley, A. D,. Fisher, M. B. and Grobicki, A. M. W. (1997). Applications of anaerobic digestion for the treatment of industrial wastewaters in Europe. *J. C. Instn Wat. Envir. Managt*, **11**, 39.

Problems

1. An anaerobic digestion plant is to give 90 per cent BOD removal to $100\,m^3$/day of slaughterhouse effluent with an ultimate BOD of $3500\,mg/l$. The solids retention time is 10 days. Calculate the daily solids accumulation and the daily gas production. Synthesis constant (Y) is 0.1 and the endogenous respiration constant (b) is 0.01/day. ($28.6\,kg$, $96\,m^3$)

2. In a treatment plant $250\,m^3$ of primary sludge are produced daily with a total solids content of 5 per cent, volatile matter is 65 per cent of the total solids. Determine the anaerobic digester capacity required for a loading of $0.75\,kg\ VS/m^3$ day and calculate the nominal retention time. ($10\,830\,m^3$, $43.3\,days$)

16

Disinfection

The small size of microorganisms means that complete removal of them from water by processes such as coagulation and filtration cannot be guaranteed. In the case of many groundwaters there may be no apparent need for treatment, but the presence of bacteria and viruses is always possible. Because of the public health significance of waterborne microorganisms it is thus essential to ensure the elimination of potentially harmful microorganisms from potable waters by the use of a suitable disinfection process. Domestic wastewater and many industrial discharges contain large numbers of microorganisms and conventional waste-water treatment processes are not primarily intended to remove pathogenic microorganisms although their numbers are significantly reduced after treatment. Where such effluents are discharged to bathing waters or are used for irrigation purposes destruction of the pathogens and indicator organisms for human pollution may be desirable. Universal disinfection of wastewaters is usually considered undesirable since the removal of most of the microorganisms will inhibit the self-purification process in the receiving water and disinfectant residuals and by-products may harm aquatic life.

It is important to differentiate between disinfection, which implies the killing of potentially harmful organisms, and sterilization, which means killing all living organisms. Potable water supplies are normally disinfected and sterilized waters are only used for medical or pharmaceutical purposes.

16.1 Theory of disinfection

In general the rate of kill is given by

$$\frac{dN}{dt} = -KN \tag{16.1}$$

where K = reaction rate constant for a particular disinfectant and N = number of viable organisms.

Integrating gives

$$\log_e \frac{N_t}{N_0} = -Kt \tag{16.2}$$

where N_0 = number of organisms initially and N_t = number of organisms at time t.
Changing to base 10,

$$\log \frac{N_t}{N_0} = -kt \tag{16.3}$$

where $k = 0.4343 \ K$, or

$$t = \frac{1}{k} \log \frac{N_0}{N_t} \tag{16.4}$$

Since N_t will never, in practice, reach zero it is usual to specify kill as a percentage, e.g. 99.9 per cent. The rate constant (K) as well as depending on the particular disinfectant also varies with disinfectant concentration, temperature, pH, the microorganism concerned and other environmental factors.

The most popular disinfectant for water is chlorine which does not obey equation 16.1 but follows the relation

$$\frac{dN}{dt} = -KNt \tag{16.5}$$

or, integrating and changing to base 10,

$$t^2 = \frac{2}{k} \log \frac{N_0}{N_t} \tag{16.6}$$

At pH 7, values of k for chlorine are about 1.6×10^{-2}/s for free residuals and 1.6 $(10^{-5}$/s for combined residuals, when applied to coliform organisms.

An alternative way of expressing the disinfecting performance of chlorine is

$$S_L c^n t = -\log \frac{N_0}{N_t} \tag{16.7}$$

where S_L = coefficient of specific lethality, c = concentration of disinfectant and n = dilution coefficient. The values of the coefficients in equation (16.7) depend upon the microorganism under consideration, the percentage kill required and the form in which the disinfectant is present. For given conditions it can be taken that

$$c^n t = \text{a constant}$$

Thus, within reason, the required degree of disinfection can be achieved by a high dose for a short contact time or a lower dose for a longer contact time. It is important to ensure that the flow patterns in the disinfection contact tank are such that the actual flow-through time is as close as possible to the theoretical plug-flow residence time. If the flow-through characteristics are unstable with significant amounts of short-circuiting the required dose and contact time combination may not be achieved, with detrimental effects on the disinfection process.

16.2 Chlorine

Chlorine (and its compounds) is widely used for the disinfection of water because it

- is readily available as gas, liquid, or powder
- is cheap
- is easy to apply due to relatively high solubility (7000 mg/l)
- leaves a residual in solution which, while not harmful to humans, provides protection in the distribution system
- is very toxic to most microorganisms, stopping metabolic activities.

It has some disadvantages in that it is a poisonous gas which requires careful handling and it can produce disinfection by-products (DBPs) which may give rise to tastes and odours and might be long-term health hazards.

Chlorine is a powerful oxidizing agent which will rapidly combine with reducing agents and unsaturated organic compounds, e.g.

$$H_2S + 4Cl_2 + 4H_2O \rightarrow H_2SO_4 + 8HCl$$

This immediate chlorine demand must be satisfied before chlorine becomes available for disinfection. 1 mg/l of chlorine will oxidize 2 mg/l BOD, but this is not normally a feasible method of wastewater treatment because of cost and the large amounts of by-products which would be produced.

After the chlorine demand has been satisfied the following reactions can occur, depending on whether ammonia is present. *In the absence of ammonia,*

$$Cl_2 + H_2O \leftrightarrow HCl + HClO$$
$$\updownarrow \quad \updownarrow \qquad \text{(free residuals)}$$
$$H^+ + Cl^- \quad H^+ + ClO^-$$

Hypochlorous acid, HClO, is the more effective disinfectant, the chlorite ion ClO^- being relatively ineffective. The dissociation of HClO is suppressed at acid pH values, the residual being all HClO at pH 5 and below, about half HClO at pH 7.5 and all ClO^- at pH 9. Thus the most effective disinfection occurs at acid pH levels, although pH adjustment purely for disinfection purposes is not common. *In the presence of ammonia,*

$$NH_4^+ + HClO \rightarrow NH_2Cl + H_2O + H^+$$
monochloramine

$$NH_2Cl + HClO \rightarrow NHCl_2 + H_2O \qquad \text{(combined residuals)}$$
dichloramine

$$NHCl_2 + HClO \rightarrow NCl_3 + H_2O$$
nitrogen trichloride

The combined residuals are more stable than free residuals but less effective as disinfectants. For a given kill with constant residual the combined form requires a hundred times the contact time required by the free residual. Alternatively, for a constant contact time the combined residual concentration must be twenty-five times the free residual concentration to give the desired kill.

Ammonia may be added to waters lacking in ammonia to allow the formation of chloramines which tend to cause less trouble with tastes and odours than do free residuals. In the presence of ammonia the continued addition of chlorine produces a characteristic residual curve as shown in Figure 16.1.

Once all the ammonia has reacted, any further chlorine converts the combined residual into a free residual at the break point, simplified as

$$NCl_3 + Cl_2 + H_2O \rightarrow HClO + NH_4^+$$

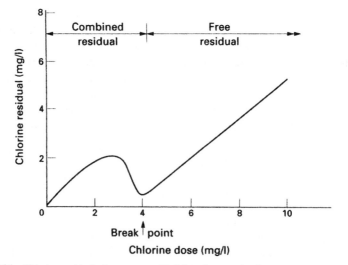

Figure 16.1 Chlorine residuals for a water with 0.5 mg/l ammonia nitrogen.

The break point theoretically occurs at 2 parts Cl_2 to 1 part NH_3 but in practice the ratio is usually nearer 10:1. Beyond the break point the free residual is proportional to the dose.

Troublesome tastes and odours can be destroyed using the oxidizing action of excess chlorine in the process known as superchlorination, the excess chlorine being removed by sulphonation, after the desired contact time, according to the reaction

$$Cl_2 + SO_2 + H_2O \rightarrow H_2SO_4 + HCl$$

Concern has recently been expressed in some circles about the presence in water of small concentrations of DBPs many of which are organochlorine compounds, some of which are carcinogenic at relatively high doses in animals. There thus may be a potential hazard in the lifetime consumption of drinking water with concentrations of a few $\mu g/l$ of these disinfection by-products. The organochlorines, of which trihalomethanes (THMs) such as chloroform are the most common, occur when raw waters containing organic matter, e.g. from sewage effluents, are disinfected with chlorine. There is no scientific evidence that the levels currently found in water supplies are in any way hazardous, but it is sensible to endeavour to prevent their formation by careful process control and avoidance of unnecessary use of chlorine. Attempts are frequently made to prevent THM formation by adjusting treatment processes to remove the precursors of these substances before chlorination is carried out.

It has recently been stressed by WHO that the risks of health hazards from DBPs are infinitesimal when compared with those due to ineffective disinfection. WHO recommends a minimum free residual of 0.5 mg/l after 30 min contact time at a pH of less than 8 and a turbidity of less than 1 NTU. Such a dose will unfortunately have no effect on *Cryptosporidium* oocysts which cannot in practice be inactivated by chlorine although *Giardia* cysts are usually inactivated by normal chlorine doses.

Additional pressures are developing in some quarters against the use of gaseous chlorine because of the potential hazard arising from a gas leak at the treatment plant or during the transport of bulk chlorine. Interest has developed into on-site electrolytic chlorine (OSEC) generation to reduce these hazards. Electrolysis of 3 per cent brine will produce a 1 per cent solution of sodium hypochlorite. For small plants hypochlorite solution provides a means of chlorination without the potential safety hazards of gaseous addition. This said, it should be noted that there is no record of serious accidents arising from the use of chlorine gas in water treatment plants.

Chlorine dioxide is mainly used for the control of tastes and odours since although there is some evidence that it is a more powerful disinfectant than chlorine in alkaline conditions, it is much more expensive. It does not combine with ammonia to any appreciable extent so that chlorine dioxide can be used to obtain free chlorine residuals in a water with large amounts of ammonia. The presence of significant amounts of ammonia in a water supply is, however, undesirable because of its nutrient value which tends to encourage biological growth. Chlorine dioxide is unstable and must be generated in situ by the action of chlorine or an acid on sodium chlorite

$$2NaClO_2 + Cl_2 \rightarrow 2ClO_2 + 2NaCl$$

$$5NaClO_2 + 4HCl \rightarrow 4ClO_2 + 5NaCl + 2H_2O$$

The possible formation of organochlorine compounds by chlorine dioxide has not yet been fully investigated, so that its use in place of chlorine may not necessarily solve the DBP problem. Chlorine dioxide has similar disinfecting power to a free chlorine residual but it is more effective at inactivating *Cryptosporidium* and *Giardia* cysts.

16.3 Ozone

Ozone (O_3) is an allotropic form of oxygen produced by passing dry oxygen or air through an electrical discharge (5000–20 000 V, 50–500 Hz). It is an unstable, highly toxic blue gas with a pungent odour of new mown hay. A powerful oxidizing agent, it is an efficient disinfectant and useful in bleaching colour and removing tastes and odours. Like oxygen it is only slightly soluble in water and because of its unstable form it leaves no residual. It is a highly toxic gas with a maximum allowable continuous concentration in air in a working environment of 0.1 p.p.m. Unless cheap energy is available ozone treatment is much more expensive than chlorination but it does have the advantage of providing good removals of colour. In these circumstances, filtration and ozonization may give a finished water similar to that produced by a more complex coagulation, sedimentation, filtration and chlorination plant. Because of the absence of ozone residuals in the distribution system, biological growths with attendant colour, taste and odour problems may result. Such growths in the distribution system can usually be prevented by adding a small dose of chlorine after ozonization and most plants using ozone also add chlorine at the final stage of the treatment process to ensure a residual in supply. Ozone has some application in the oxidation of certain industrial wastewaters not amenable to biological oxidation.

Ozone must be manufactured on site by passing dry air through a high-voltage high-frequency electrical discharge. There are two main types of ozonizer: plate type with flat electrodes and glass dielectrics, and the tube type with cylindrical electrodes coaxial with glass dielectric cylinders. The high-tension side is cooled by convection and the low tension side by water. Air passes between the electrodes and is ozonized by the discharge across the air gap. Ozone production is usually up to about 4 per cent by weight of the carrier air with power requirements of around 25 kW h/kg of ozone produced. Because of the limited solubility of ozone (about 30 mg/l) and its highly reactive nature, rapid mixing of the gas with the water is essential. This is usually achieved with a fine bubble diffuser placed in the bottom of a contact chamber. Any excess ozone escaping from the chamber may be collected and recycled to save energy or destroyed to prevent a hazard to workers on site. Ozone will react with organic matter to form ozonides in certain conditions and the significance of the presence of these products in water is not yet fully understood. Any bromine in the water will be converted by ozonization to bromate which is potentially a long-term health risk. Ozone is a more powerful disinfectant than chlorine and a typical requirement is 0.4 mg/l residual after 4 min contact. *Cryptosporidium* oocysts are inactivated at a 2 mg/l ozone after 10 min contact and *Giardia* cysts are inactivated after 1 min at 0.5 mg/l.

16.4 Ultraviolet radiation

Various forms of radiation can be effective disinfecting agents and UV radiation has been used for the treatment of small water supplies for many years. The disinfecting action of UV at a wavelength of around 254 nm is quite strong provided that the organisms are actually exposed to the radiation. It is thus necessary to ensure that turbidity is absent and that the dose is increased to allow for the absorption of UV by any organic compounds present in the flow. The water to be disinfected flows between mercury arc discharge tubes and polished metal reflector tubes which gives efficient disinfection with a retention time of a few seconds although at a rather high power requirement of 10–20 W/m^3 h. Provided the water has a turbidity below 1 NTU a UV dose of 15–25 mW s/cm^2 is usually sufficient to give a 99.9 per cent kill of most microorganisms. UV radiation has no effect on *Cryptosporidium* and *Giardia* cysts at normal doses. The advantages of UV disinfection include no formation of tastes and odours, minimum maintenance, easy automatic control with no danger from overdosing. Disadvantages are lack of residual, high cost and need for high clarity in the water. Because of the absence of DBPs, UV treatment is favoured by regulatory bodies for the disinfection of wastewater effluents where this is required because of bathing water standards or where water-contact sports are undertaken.

16.5 Other disinfectants

Various other methods have been used for water disinfection, including those listed below.

1. *Heat*. Disinfection by heat is very effective but is costly and impairs the palatability of water by removing DO and dissolved salts. There is no residual effect.

2. *Silver.* Colloidal silver was used by the Romans to preserve the quality of water in storage jars since, at concentrations of about 0.05 mg/l, silver is toxic to most microorganisms. It is of value for small portable filter units for field use where silver-impregnated gravel filter candles remove turbidity and provide disinfection. The cost becomes excessive for other than very small supplies.
3. *Bromine.* A halogen like chlorine, bromine has similar disinfection properties and is sometimes used in swimming pools where the residual tends to be less irritating to the eyes than chlorine residuals.
4. *Iodine.* Another halogen, which is occasionally used for disinfection of swimming pools because again the residuals are less irritating than those of chlorine. It is sometimes used for the disinfection of personal water supplies when travelling in remote areas but is not used for public water supplies.
5. *Membranes.* Pressure-driven membrane systems such as microfiltration and ultrafiltration are capable of removing particles in the range from 5 μm down to 10^{-2} μm which covers the size of most microorganisms of importance in water quality.

Further reading

Brett, R. W. and Ridgway, J. W. (1981). Experiences with chlorine dioxide in Southern Water Authority and Water Research Centre. *J. Instn Wat. Engnrs Scientists*, **35**, 135.

Cross, T. S. C. and Murphy, R. (1993). Disinfection of sewage effluent: The Jersey experience. *J. Instn Wat. Envir. Managt*, **7**, 481.

Edge, J. C. and Finch, P. E. (1987). Observations on bacterial aftergrowth in water supply distribution systems: implications for disinfection strategies. *J. Instn Wat. Envir. Managt*, **1**, 104.

Falconer. R. A. and Tebbutt, T. H. Y. (1986). Theoretical and hydraulic model study of a chlorine contact tank. *Proc. Instn Civ. Engnrs*, **81**, 255.

Fawell, J. K., Fielding, M. and Ridgway, J. W. (1987). Health risks of chlorination—is there a problem? *J. Instn Wat. Envir. Managt*, **1**, 61.

Ferguson, D. W., Gramith, J. T. and McGuire, M. J. (1991). Applying ozone for organics control and disinfection: A utility perspective. *J. Amer. Wat. Wks Assn*, **83**(5), 32.

Gordon, G., Cooper, W. J., Rice, R. G. and Pacey, G. E. (1988). Methods of measuring disinfectant residuals. *J. Am. Wat. Wks Assn*, **80**(9), 94.

Job, G. D., Trengrove, R. and Realey, G. J. (1995). Trials using a mobile ultraviolet disinfection system in South West Water. *J. C. Instn Wat. Envir. Managt*, **9**, 257.

McGuire, M. J. (1989). Preparing for the disinfection by-products rule: A water industry status report. *J. Am. Wat. Wks Assn*, **81**(8), 35.

Myers, A. G. (1990). Evaluating alternative disinfectants for THM control in small systems. *J. Am. Wat. Wks Assn*, **82**(6), 77.

Palin, A. T. (1980). Disinfection. In *Developments in Water Treatment*, Vol. 2 (W. M. Lewis, ed.). Barking: Applied Science Publishers.

Rudd, T. and Hopkinson, L. M. (1989). Comparison of disinfection techniques for sewage and sewage effluents. *J. Instn Wat. Envir. Managt*, **3**, 612.

Smith, D. J. (1990). The evolution of an ozone process at Littleton water treatment works. *J. Instn Wat. Envir. Managt*, **4**, 361.

Stevenson, D. G. (1995). The design of disinfection contact tanks. *J. C. Instn Wat. Envir. Managt*, **9**, 146.

Symons, J. M., Krasner, S. W., Simms, L. A. and Selimenti, M. (1993). Measurement of THM and precursor concentrations revisited: The effect of bromide ion. *J. Amer. Wat. Wks Assn*, **85**(1), 51.

Tetlow, J. A. and Hayes, C. R. (1988). Chlorination and drinking water quality—an operational overview. *J. Instn Wat. Envir. Managt*, **2**, 399.

Thomas, V. K., Realey, G. J. and Harrington, D. W. (1990). Disinfection of sewage, stormwater and final effluent. *J. Instn Wat. Envir. Managt*, **4**, 422.

World Health Organization (1989). *Disinfection of Rural and Small-Community Water Supplies*. Medmenham: WRC.

Problems

1. Ozone is to be used to obtain a 99.9 per cent kill of bacteria in water with a residual of 0.5 mg/l. Under these conditions the reaction constant (k) is 2.5×10^2/s. Determine the contact time required. (120 s)

2. Compare the contact times necessary to give *E. coli* kills of 99.99 per cent in water with (a) free chlorine residual of 0.2 mg/l, and (b) combined chlorine residual of 1 mg/l. k values are 10^{-2}/s and 10^{-5}/s respectively. (28 s, 890 s)

17

Chemical treatment

A number of constituents of waters and wastewaters do not respond to the conventional treatment processes already discussed, and alternative forms of treatment must be used for their removal. Soluble inorganic matter can often be removed by precipitation or ion-exchange techniques. Soluble non-biodegradable organic substances may frequently be removable by adsorption.

17.1 Chemical precipitation

Removal of certain soluble inorganic materials can be achieved by the addition of suitable reagents to convert the soluble impurities into insoluble precipitates which can then be flocculated and removed by sedimentation. The extent of removal which can be accomplished depends on the solubility of the product; this is usually affected by such factors as pH and temperature.

Chemical precipitation may be used in industrial wastewater treatment, for example to remove toxic metals from metal-finishing effluents. Such effluents often contain considerable amounts of hexavalent chromium which is harmful to biological systems. By the addition of ferrous sulphate and lime the chromium is reduced to the trivalent form which can be precipitated as a hydroxide as shown below

$$Cr^{6+} + 3Fe^{2+} \rightarrow Cr^{3+} + 3Fe^{3+}$$

$$Cr^{3+} + 3OH^- \rightarrow \underline{Cr(OH)_3}$$

$$Fe^{3+} + 3OH^- \rightarrow \underline{Fe(OH)_3}$$

For efficient treatment it is essential to add the correct dose of reagent. For chromium reduction the theoretical requirement is shown above, but at this level the reaction proceeds very slowly and in practice, to ensure complete reduction, it is necessary to add 5–6 atoms of ferrous iron for each atom of hexavalent chromium. Phosphates can be precipitated from solution by the addition of metal ions, a process used for the removal of nutrients from wastewater effluents and described in Chapter 19. A characteristic of chemical precipitation processes is the production of relatively large volumes of sludge.

A common use of chemical precipitation is in water softening. Hard waters, i.e. waters containing calcium and magnesium in significant amounts, often require softening to improve their suitability for washing and heating purposes. For potable supplies water with up to about 75 mg/l hardness (as $CaCO_3$) is usually considered as soft, but some surfacewaters and many groundwaters can have hardness levels of several hundred mg/l. A hardness in excess of 300 mg/l would normally be considered undesirable. The need for softening of domestic supplies depends on reasons of convenience and economy rather than of health, since even at very high concentrations of up to 1000 mg/l hardness is quite harmless. Indeed there is some statistical evidence to suggest that artificially softened waters may increase the incidence of some forms of heart disease.

Hardness is normally expressed in terms of calcium carbonate whereas chemical analyses for individual ions are usually given in terms of that ion. It is thus necessary to convert the analytical results to the common denominator

$$\text{X (mg/l as } CaCO_3) = \text{X (mg/l)} \times \frac{\text{equivalent weight } CaCO_3}{\text{equivalent weight X}} \quad (17.1)$$

where X is any ion or radical.

Thus for a typical water analysis conversion to concentrations in terms of $CaCO_3$ may be carried out as follows

$$40 \text{ mg/l } Ca^{2+} \times 50/20.04 = 99.0 \text{ mg/l as } CaCO_3$$

$$24 \text{ mg/l } Mg^{2+} \times 50/12.16 = 98.5 \text{ mg/l as } CaCO_3$$

$$9.2 \text{ mg/l } Na^+ \times 50/23 = 20.0 \text{ mg/l as } CaCO_3$$

$$183 \text{ mg/l } HCO_3^- \times 50/61 = 150.0 \text{ mg/l as } CaCO_3$$

$$57.5 \text{ mg/l } SO_4^{2-} \times 50/48 = 58.0 \text{ mg/l as } CaCO_3$$

$$7.0 \text{ mg/l } Cl^- \times 50/35.5 = 9.5 \text{ mg/l as } CaCO_3$$

Note that the cations and anions both sum to 217.5 mg/l. It is now possible to represent the composition of the water as a bar diagram (Figure 17.1).

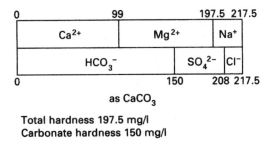

Total hardness 197.5 mg/l
Carbonate hardness 150 mg/l

Figure 17.1 Composition of a water sample in terms of $CaCO_3$.

Precipitation softening is based on the reversal of the process by which the hardness entered the water initially, i.e. the conversion of soluble compounds into insoluble ones which will then precipitate and permit removal by flocculation and sedimentation. The method of precipitation softening adopted depends on the form of hardness present.

Lime softening

For calcium hardness of the carbonate form, the addition of lime equivalent to the amount of bicarbonate present will form insoluble calcium carbonate

$$Ca(HCO_3)_2 + Ca(OH)_2 \rightarrow \underline{2CaCO_3} + 2H_2O$$

The solubility of $CaCO_3$ at normal temperatures is about 20 mg/l and because of the limited contact time available in a treatment plant a residual of about 40 mg/l $CaCO_3$ usually results. The softened water is thus saturated with $CaCO_3$ and deposition of scale would be likely in the distribution system; this may be prevented by carbonation which produces soluble $Ca(HCO_3)_2$

$$CaCO_3 + CO_2 + H_2O \rightarrow Ca(HCO_3)_2$$

or by the addition of phosphates which sequester calcium and prevent scaling. Figure 17.2 shows the steps in lime softening.

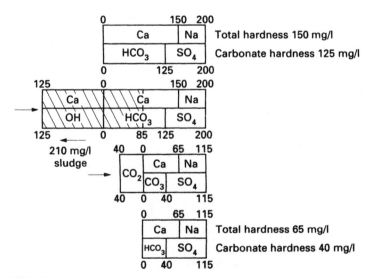

Figure 17.2 Lime softening.

Lime-soda softening

This is used for all forms of calcium hardness. By adding soda ash (Na_2CO_3), non-carbonate hardness is converted to $CaCO_3$ which will then precipitate

$$CaSO_4 + Na_2CO_3 \rightarrow CaCO_3 + Na_2SO_4$$

As much soda ash is added as there is non-carbonate hardness associated with calcium (Figure 17.3).

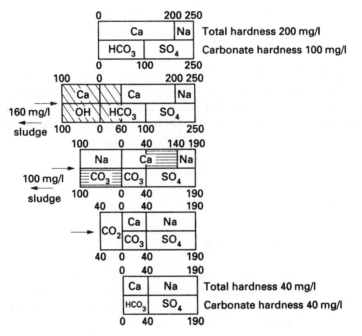

Figure 17.3 Lime-soda softening.

Excess-lime softening

This is used for magnesium carbonate hardness. The above methods are not effective for the removal of magnesium because magnesium carbonate is soluble

$$Mg(HCO_3)_2 + Ca(OH)_2 \rightarrow \underline{CaCO_3} + MgCO_3 + 2H_2O$$

However, at about pH 11

$$MgCO_3 + Ca(OH)_2 \rightarrow \underline{Mg(OH)_2} + \underline{CaCO_3}$$

The practical solubility of $Mg(OH)_2$ is about 10 mg/l. For excess lime softening it is necessary to add

$$Ca(OH)_2 \equiv HCO_3^-$$

$$+ Ca(OH)_2 \equiv Mg^{2+}$$

$$+ 50 \text{ mg/l excess } Ca(OH)_2 \text{ to raise pH to } 11$$

The high pH level produces good disinfection as a by-product and thus chlorination might be unnecessary after such softening except for a small dose to provide a residual chlorine in the distribution system. Carbonation is necessary to remove the excess lime and reduce the pH after treatment.

Excess-lime soda softening

This is used for all forms of magnesium hardness and involves the use of lime and soda ash and is a complicated process.

All forms of precipitation softening produce considerable volumes of sludge. Lime recovery is possible by calcining $CaCO_3$ sludge and slaking with water

$$CaCO_3 \rightarrow CaO + CO_2$$

$$CaO + H_2O \rightarrow Ca(OH)_2$$

In this way more lime than is required in the plant is produced and the surplus may be sold, the sludge-disposal problem having been solved at the same time.

Precipitation softening is sometimes carried out in upflow reactors which are seeded with calcium carbonate pellets which provide nuclei for deposition of more carbonate from solution. As the pellets grow in size they sink to the bottom of the reactor vessel from where they can be removed as a readily draining granular material suitable for agricultural use.

17.2 Ion exchange

Certain natural materials, notably zeolites which are complex sodium alumino-silicates and greensands, have the property of exchanging one ion in their structure for another ion in solution. Synthetic ion-exchange materials have been developed to provide higher exchange capacities than the natural compounds.

Ion-exchange treatment has the advantage that no sludge is produced, but it must be remembered that when the ion-exchange capacity has been exhausted the material must be regenerated, which gives rise to a concentrated waste stream of the original contaminant. Industrial wastewaters, such as metal-finishing effluents, can be treated by ion exchange as an alternative to precipitation methods and nitrates can be removed from drinking waters. The commonest use of ion exchange, however, is for water softening or demineralization in the case of high-pressure boiler feed waters where high-purity water is essential.

When used for water softening, natural zeolites will exchange their sodium ions for the calcium and magnesium ions in the water, thus giving complete removal of hardness, i.e. (representing a zeolite by Na_2X)

$$\left.\begin{array}{c} Ca^{2+} \\ Mg^{2+} \end{array}\right\} + Na_2X \rightarrow \left.\begin{array}{c} Ca \\ Mg \end{array}\right\} X + 2Na^+$$

The finished water is thus high in sodium, which is not likely to be troublesome unless the water was originally very hard. When all sodium ions in the structure have been exchanged, no further removal of hardness occurs. Regeneration can then be carried out using a salt solution to provide a high concentration of sodium ions to reverse the exchange reaction, the hardness being released as a concentrated chloride stream.

$$\left.\begin{array}{c} Ca \\ Mg \end{array}\right\} X + 2NaCl \rightarrow Na_2X + \left.\begin{array}{c} Ca \\ Mg \end{array}\right\} Cl_2$$

A natural sodium cycle zeolite will have an exchange capacity of about 200 gram equivalents/m^3 with a regenerant requirement of about 5 equivalents/ equivalent exchanged. Synthetic sodium cycle resins may have double the exchange capacity with about half the regenerant requirement, but have a higher capital cost.

Hydrogen cycle cation exchangers produced from natural or synthetic carbonaceous compounds are available and will also give a water of zero hardness. They exchange all cations for hydrogen ions so that the product stream is acidic and their main use is as the first stage in demineralization operations.

$$\left.\begin{array}{c} Ca^{2+} \\ Mg^{2+} \\ 2Na^+ \end{array}\right\} + H_2Z \rightarrow \left.\begin{array}{c} Ca \\ Mg \\ 2NA \end{array}\right\} Z + 2H^+$$

Regeneration is by acid treatment

$$\left.\begin{array}{c} Ca \\ Mg \\ 2NA \end{array}\right\} Z + 2H^+ \rightarrow H_2Z + \left\{\begin{array}{c} Ca^{2+} \\ Mg^{2+} \\ 2Na^+ \end{array}\right.$$

Typical performance characteristics for a hydrogen cycle exchanger are 1000 gram equivalents/m^3 exchange capacity and a regenerant requirement of about 3 equivalents/equivalent exchanged. Anion exchangers which are usually synthetic ammonia derivatives will accept the product water from a hydrogen cycle exchanger and produce demineralized water for laboratory and other specialized uses as well as for boiler feed water.

A strong anion exchanger ROH, where R represents the organic structure, will remove all anions

$$\left.\begin{array}{l} HNO_3 \\ H_2SO_4 \\ HCl \\ H_2SiO_3 \\ H_2CO_3 \end{array}\right\} + ROH \rightarrow R\left\{\begin{array}{l} NO_3 \\ SO_4 \\ Cl \\ SiO_3 \\ CO_3 \end{array}\right. + H_2O$$

A strong base is necessary for regeneration

$$R\left\{\begin{array}{l} NO_3 \\ SO_4 \\ Cl \\ SiO_3 \\ CO_3 \end{array}\right. + ROH \rightarrow R\left\{\begin{array}{l} NO_3 \\ SO_4 \\ Cl \\ SiO_3 \\ CO_3 \end{array}\right. + H_2O$$

Weak anion exchangers remove strong anions but not carbonates and silicates.

Typical anion exchanger performance would be $800\,\text{gram equivalents/m}^3$ exchange capacity and 6 equivalents/equivalent exchanged regenerant capacity.

Ion-exchange materials are normally used in units similar to pressure filters and it is possible to combine cation and anion resins in a single mixed-bed unit. The feedwater to an ion-exchange plant should be free of suspended matter, since this would tend to coat the surfaces of the exchange medium and reduce its efficiency. Organic matter in the feed can also cause fouling of the exchanger, although the development of macroporous materials with their internal surface area inaccessible to large organic molecules reduces the problems of organic fouling.

The growing concern about the presence of nitrates in raw waters, particularly those from underground sources, has focused attention on nitrate removal processes. A major problem lies in the high solubility of all nitrates so that precipitation is not a viable removal process. By blending a high-nitrate source with another source containing little or no nitrate an acceptable quality can be obtained for distribution. Where blending is not available it is possible to use ion-exchange techniques to remove nitrates since they occur as anions in solution. Conventional anion exchangers are not selective for nitrate ions and thus their exchange capacity may largely be utilized by unnecessary removal of other anions such as sulphate. New nitrate-specific exchange resins are now available and a number of plants have been commissioned. They do of course produce a high nitrate content waste stream which also has a high chloride concentration so that disposal of the waste from regeneration must be controlled carefully.

17.3 Adsorption

Biological treatment processes can be very effective for removing organic contaminants from wastewaters, but since they rely on the organic matter providing food for the microorganisms they cannot be expected to remove small concentrations of organics. Nor can such processes remove non-biodegradable organic substances, which are not uncommon in certain industrial wastewaters. It has long been recognized that tiny concentrations of both natural and synthetic organic substances can in some cases produce serious taste and odour problems in potable waters. More recently the ability to determine nanogram concentrations of individual organic compounds coupled with increasing information relating to the potential health hazards of long-term exposure to such substances has focused more attention on their removal. The EC drinking water directive MACs of 0.5 μg/l for total pesticides and 0.1 μg/l for individual pesticides have generated increased interest in the use of adsorption techniques to remove soluble organics from drinking waters. Adsorption is in general seen as more acceptable than chemical oxidation processes using chlorine or ozone which tend to leave oxidation by-products in solution which in turn may themselves have possible health risks.

Adsorption is the accumulation of molecules from a substance dissolved in a solvent on to the surface of an adsorbent particle. The solute molecule is held in contact with the adsorbent by a combination of physical, ionic and chemical forces. Because adsorption is a surface phenomenon, effective adsorbents have a highly porous structure so that their surface area to volume ratio is very large. It is possible to custom make an adsorbent so that its internal pore dimensions are related to the molecular size of the contaminant to be removed.

When an adsorbent is left in contact with a solution the amount of adsorbed solute increases on the surface of the adsorbent and decreases in the solvent. After some time an adsorption equilibrium is reached when the number of molecules leaving the surface of the adsorbent is equal to the number of molecules being adsorbed on the surface. The nature of the adsorption reaction can be described by relating the adsorption capacity (mass of solute adsorbed per unit mass of absorbent) to the equilibrium concentration of solute remaining in solution. Such a relation is known as an adsorption isotherm (Figure 17.4).

Two simple mathematical models of adsorption are available. The Langmuir isotherm was developed from a theoretical consideration of adsorption based on the concept of equilibrium in a monomolecular surface layer

$$\frac{x}{m} = \frac{abc}{1 + ac} \tag{17.2}$$

where x = mass of solute adsorbed, m = mass of adsorbent, c = solute concentration remaining at equilibrium and a and b are constants depending on

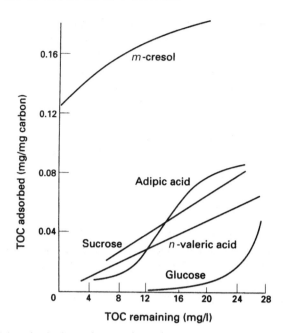

Figure 17.4 Adsorption isotherms for several organics.

adsorbent, solute and temperature. The Freundlich equation is an empirical relationship which often gives a more satisfactory model for experimental data

$$\frac{x}{m} = kc^{1/n} \tag{17.3}$$

where k and n are constants depending upon adsorbent, solute and temperature.

In general, the overall rate of adsorption is governed by the rate of diffusion of the solute into the capillary pores of the adsorbent particle. The rate decreases with increasing particle size and increases with increasing solute concentration and increasing temperature. High molecular weight solutes are not adsorbed as readily as low molecular weight substances and increasing solubility decreases the adsorbability of organic compounds.

The most popular adsorbent for many uses is activated carbon which is produced from coal, wood or vegetable fibre sources. Dehydration and carbonization are achieved by slow heating in the absence of air and the activation effect is obtained by the application of steam, air or carbon dioxide at temperatures of around 950°C. Activated carbon may be used in the form of powder (PAC), with a particle size of 10–50 μm, added to sedimentation tanks or to the surfaces of sand filters at doses of 10–40 mg/l. This technique is useful where there is intermittent need of adsorption to remove occasional traces of organics. When regular removal is required it is more effective to use granular carbon (GAC) with a particle size of 0.5–2 mm and a surface area of around 1000 m^2/g. The GAC is usually placed in downflow beds similar to conventional sand filters and they may replace the sand filtration stage or, with turbid waters, be placed after the sand filtration stage. Carbon adsorber beds are usually designed using the parameter 'empty bed contact time' (EBCT) which is the retention time of the contactor in the absence of the carbon. Typical values of EBCT for potable water treatment are 10–20 min. The term 'effective carbon dose' (ECD) is sometimes used to express plant performance and is defined as

$$\text{ECD} = \frac{\text{weight of GAC in bed (g)}}{\text{volume of water treated during service (m}^3\text{)}} \qquad (17.4)$$

Use of GAC as a filter bed is likely to reduce the adsorptive capacity as turbidity particles cover the carbon surfaces. As mentioned in Chapter 13, Thames Water has developed a sandwich slow sand filter which has a 125 mm layer of GAC in the middle of the 900 mm bed.

When the adsorptive capacity of the carbon has been exhausted the carbon must be discarded if in PAC form or regenerated if in the more expensive GAC form. Regeneration is normally achieved by firing in a furnace under the conditions described earlier. There is some loss of adsorption capacity after regeneration and the physical handling of the GAC causes some mechanical attrition with further loss of active carbon. It is thus necessary to add 5–10 per cent of fresh or virgin carbon to the regenerated material so that its adsorption capacity per unit weight is fully restored. The adsorbed organic substances are released in the regeneration furnace which must be provided with effective combustion and emission controls to prevent environmental contamination. Large treatment plants using GAC may have on-site regeneration facilities, but in most cases a replacement and regeneration service is offered by the suppliers of the carbon.

Further reading

Adams, J. Q. and Clark, R. M. (1989). Cost estimates for GAC treatment systems. *J. Am. Wat. Wks Assn*, **81**(1), 30.

Andrews, D. A. and Harward, C. (1994). Isleham ion-exchange nitrate removal plant, *J. Instn Wat. Envir. Managt*, **8**, 120.

Hyde, R. A. (1989). Application of granular activated carbon in the water industry. *J. Instn Wat. Envir. Managt*, **3**, 174.

Martin, R. J. and Iwugo, K. O. (1979). Studies on residual organics in biological plant effluents and their treatment by the activated carbon adsorption process. *Pub. Hlth Engnr*, **7**, 61.

Najim, I. M., Soenyink, V. L., Lykins, B. W. and Adams, J. Q. (1991). Using powdered activated carbon: A critical review. *J. Am. Wat. Wks Assn*, **83**(1), 65.

Short, C. S. (1980). Removal of organic compounds. In *Developments in Water Treatment*, Vol. 2 (W. M. Lewis, ed.). Barking: Applied Science Publishers.

Problems

1. Make up a bar diagram in terms of calcium carbonate for a water with the following composition

Ca^{2+}	101.0 mg/l
Mg^{2+}	4.75 mg/l
Na^+	14.0 mg/l
HCO_3^-	220.0 mg/l
SO_4^{2-}	88.4 mg/l
Cl^-	21.3 mg/l

Ca 40, Mg 24.3, Na 23, H 1, C 12, O 16, S 32, Cl 35.5. (Total hardness 271.6 mg/l, carbonate hardness 180 mg/l)

2. Soften the water in question 1 by lime and lime-soda treatment and determine the final hardness of the water in each case. (Total hardness 131.6 mg/l and 59.6 mg/l. Carbonate hardness 40 mg/l in both cases)

3. A sodium cycle cation exchanger has a volume of 10 m³ and an exchange capacity of 400 gram equivalents/m³. Determine the volume of water of initial hardness 250 mg/l as $CaCO_3$ which can be softened in the exchanger. If regeneration requires 5 equivalents/equivalent exchanged, determine the amount of sodium chloride necessary for regeneration. (800 m³, 1.17 tonnes)

4. The following data were obtained from laboratory tests using powdered activated carbon:

Carbon dose (mg/l)	Initial TOC (mg/l)	Final TOC (mg/l)
12	20	5
7	12	2

Determine the carbon dose necessary to reduce an initial TOC of 15 mg/l to 3 mg/l assuming that the Langmuir isotherm applies. (9.1 mg/l)

5. A water treatment plant uses GAC to remove pesticides. The flow is 10 Ml/day and the design EBCT for the carbon contact beds is 15 min. Determine the bed volume required. If the service life of the GAC is 250 days and the ECD is 30 mg/l calculate the weight of GAC in the bed. (104 m³, 75 tonnes)

18

Sludge dewatering and disposal

One of the major problems in wastewater treatment is that of sludge disposal. Large volumes of putrescible organic sludge with high water contents are produced from primary and secondary sedimentation tanks and their dewatering and ultimate disposal may account for as much as 40 per cent of the cost of treatment. Volumes of sludge from water treatment operations are much lower and their relatively inert nature means that disposal is normally less of a problem

18.1 Sludge characteristics

The types of sludge which are produced in treatment processes are

- primary sludge from wastewater sedimentation
- secondary sludge from biological wastewater treatment
- digested forms of the above separately or mixed
- hydroxide sludges from coagulation and sedimentation of waters and industrial wastes
- precipitation sludges from softening, phosphate removal and from industrial waste treatment.

All of these sludges have low solids contents (1–6 per cent) and thus large volumes of sludge must be handled to dispose of even a relatively small mass of solids. The main objective in treatment of sludge is therefore to concentrate the solids by removing as much water as possible. The density and nature of the solid particles have a considerable influence on the thickness of the sludge produced. Thus a metallurgical ore slurry with solids of relative density 2.5 will quickly separate to give a sludge with a solids content of 50 per cent or more. On the other hand, sewage sludge contains highly compressible solids with relative density of about 1.4 and will only produce a sludge of 2–6 per cent solids. Unless care is exercised, attempts to increase the solids content by draining off excess water may cause the solids to compress, thus blocking the voids and preventing further drainage. Primary sewage sludge has a heterogeneous nature with fibrous solids so that drainage is easier than from activated and digested sludges which are much more homogeneous in nature.

The relative density of a sludge with a particular solids content can be determined from the following expression

$$\text{relative density of sludge} = 100 \bigg/ \left[\left(\frac{\% \text{ solids}}{\text{RD of solids}} \right) + \left(\frac{\% \text{ water}}{1} \right) \right] \quad (18.1)$$

The influence of a reduction in moisture content on the volume occupied by a sludge can be shown by a simple example.

Worked example on sludge volumes

One tonne of a 2 per cent sludge whose solids have a relative density of 1.4 is to be dewatered to 25 per cent solids content. Determine the initial volume and the volume when dewatered to 25 per cent solids.

At 2 per cent solids one tonne contains 20 kg solids and 980 kg water and

$$\text{relative density, } d = 100/[(2/1.4) + (98/1)] = 1.006$$

$$\text{Volume} = 1000/(1.006 \times 1000) = 0.994 \, \text{m}^3$$

At 25 per cent solids the 20 kg solids is now associated with 60 kg water and

$$d = 100/[(25/1.4) + (75/1)] = 1.076$$

$$\text{Volume} = 80/(1.076 \times 1000) = 0.074 \, \text{m}^3$$

Thus by reducing the moisture content from 98 per cent to 75 per cent the volume of sludge has been reduced to 7.4 per cent of the original volume, a very significant change when considering disposal options.

The work of Carman on filtration, as stated in equation 13.3, was developed by Coackley (1955) for the dewatering of sludges by filtration. The parameter specific resistance to filtration was introduced to compare the filtrability of different sludges. Measurement involves placing a sludge sample in a standard filter and noting the liquid removed by a standard vacuum over a range of time.

The rate of filtration, i.e. the ease of dewatering, is given by

$$\frac{dV}{dt} = \frac{PA^2}{\mu(rcV + RA)} \quad (18.2)$$

where V = filtrate volume obtained after time t,
$\quad P$ = applied pressure,
$\quad A$ = filter area,
$\quad \mu$ = absolute viscosity of filtrate,
$\quad r$ = specific resistance of sludge,
$\quad c$ = solids concentration of sludge and
$\quad R$ = resistance of clean filter medium.

For constant P integration gives

$$t = \frac{\mu r c}{2PA^2} V^2 + \frac{\mu R}{PA} V \qquad (18.3)$$

or

$$\frac{t}{V} = \frac{\mu r c}{2PA^2} V + \frac{\mu R}{PA} \qquad (18.4)$$

Using a laboratory filtration apparatus it is possible to determine the specific resistance by plotting t/V against V, the slope of the line being $\mu r c/2PA^2$. The higher the value of the specific resistance the more difficult the sludge will be to dewater.

An alternative method of assessing the filtrability of a sludge is to measure the capillary suction time (CST) devised by Baskerville and Gale (1968). The CST depends on the suction applied to a sample of sludge by an absorbent chromatography paper. An area of paper in the centre is exposed to the sludge whilst the remaining paper is used to absorb the water removed from the sludge by capillary suction. The time taken for water to travel a standard distance through the paper is noted visually or electronically and is found to have good correlation with specific resistance values for a particular sludge. The longer the time for water to be absorbed from the sludge the more difficult it will be to dewater. The CST determination is much easier and quicker to perform than the specific resistance to filtration measurement.

18.2 Sludge conditioning

To improve the efficiency of the dewatering process it is often useful to provide a preliminary conditioning stage to release as much bound water as possible from the sludge particles, to encourage solids agglomeration and increase the solids content. Various methods of conditioning are employed, depending upon the characteristics of the sludge to be treated.

Thickening

With many flocculent sludges, particularly surplus activated sludge, slow-speed stirring in a tank with a picket-fence type mechanism encourages further flocculation and can significantly increase the solids content and settleability, allowing supernatant to be drawn off.

Chemical conditioning

Chemical coagulants can be useful in promoting agglomeration of floc particles and releasing water. Common coagulants used for sludge conditioning are aluminium sulphate, aluminium chlorohydrate, iron salts, lime and/or poly-electrolytes. The cost of these reagents is usually more than covered by the increased solids content and improvement in dewatering characteristics arising from their use.

Elutriation

The chemical requirement for conditioning can be reduced by mixing the sludge with water or effluent and allowing settlement and removal of supernatant to take place before chemicals are added. This washing process removes much of the alkalinity which in digested sludges exerts a high chemical demand.

Heat treatment

A number of processes have been employed to heat wastewater sludges under pressure with the aim of stabilizing the organic matter and improving dewaterability. A typical operation involves heating to a temperature of about 190°C for 30 min at a pressure of 1.5 MPa, the sludge being then transferred to thickening tanks. The supernatant has a high soluble organic content and must be returned to the main oxidation plant for stabilization, which is not always easy due to its limited biodegradability. Corrosion problems and high energy costs have meant that heat treatment plants are not now viable options in most situations.

18.3 Sludge dewatering

For many sludge-disposal methods preliminary dewatering is essential if the costs of disposal are to be kept under control and a variety of dewatering methods are employed (Figure 18.1) depending upon land availability and the costs related to the particular situation.

Drying beds

The oldest and simplest dewatering process uses shallow rectangular beds with porous bottoms above an underdrain layout. The beds are divided into convenient areas by low walls. Sludge is run on to the beds to give a depth of 125–250 mm and dewatering takes place due to drainage from the lower layers and evaporation from the surface under the action of sun and wind. The cake cracks as it dries, allowing further evaporation and the escape of rainwater from the surface. In good conditions a solids content of around 25 per cent can be achieved in a few weeks but a more normal period in temperate climates would be two months. Best results are obtained by applying shallow layers of sludge frequently rather

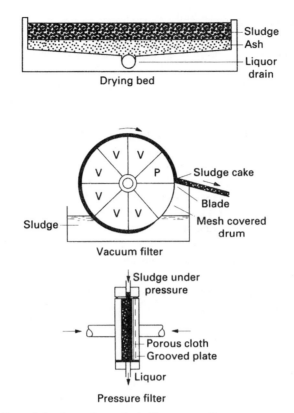

Figure 18.1 Some sludge dewatering methods. V = vacuum; P = pressure.

than deep layers at longer intervals. Removal of dried sludge can be undertaken manually at small works, but elsewhere mechanical sludge-lifting plant must be used. A typical area requirement for sewage sludge is $0.25\ m^2$/head of population and it is this large requirement which makes the feasibility of drying beds dubious unless land is available at low cost.

Mechanical dewatering devices which produce a dry friable product have grown in popularity in recent years, partly because of their small size and also because they are not affected by adverse weather conditions.

Pressure filtration

Plate or membrane filtration is a batch process in which conditioned sludge is pumped with increasing pressure into chambers lined with filter cloths or membranes which retain the solids but allow liquid to escape via grooves in the metal backing plates. As liquid escapes, the cake formed adjacent to the cloth or membrane acts as a further filter for the remainder of the sludge so that the cake dewaters towards the centre. Pressing times vary from 2 to 18 h with pressures of 600–850 kPa giving a resultant cake solids content of 25–50 per cent. Solids loading depends upon the nature of the sludge and length of pressing cycle.

A filter belt press provides continuous operation by introducing conditioned sludge via a gravity or vacuum-assisted drainage section into the gap between two endless belts to which pressure is applied by means of rollers. Dewatering occurs by a combination of gravity drainage, pressure filtration and shear.

Vacuum filtration

This is a continuous process, common in some process industries, in which a revolving segmented drum covered with filter cloth is partially submerged in conditioned sludge. A vacuum of about 90 kPa is applied to the submerged segments and sludge is attracted to the surface of the cloth. As the drum rotates and the layer of sludge emerges from the tank, air is drawn through it by the vacuum to assist dewatering. A scraper blade removes the sludge cake assisted by change to positive pressure in the relevant drum segment. The cake solids are usually 20–25 per cent with filter yields of around 20 kg dry solids/m^2 h.

Centrifugation

Continuously operated centrifuges have some application for sludge dewatering. Most are of the solid-bowl type in which conditioned sludge is fed into the centre of a rapidly rotating bowl. The solids are thrown to the outer edge of the bowl from where they are removed by a scraper/conveyor. Centrifuges are relatively compact but are not usually able to achieve solids concentrations greater than 20 per cent and in some cases it is difficult economically to reach solids contents higher than about 12–15 per cent from water and wastewater sludges. The abrasive nature of some wastewater sludges can lead to rapid wear on moving parts in centrifuges with high maintenance costs.

Sludge liquors

It is important to appreciate that in all sludge dewatering operations the separated liquid requires suitable disposal arrangements. With wastewater sludges the liquid is usually highly polluting and must be returned to the main treatment plant for stabilization where it can contribute significantly to the incoming organic load, particularly in relation to ammonia concentration and refractory organic substances. The organic load in returned sludge liquors may be as much as 10–15 per cent of the incoming load and in works which import sludge from smaller plants for central handling the returned liquor can be a significantly larger percentage of the load from the incoming wastewater.

18.4 Sludge stabilization

After dewatering to a cake, wastewater sludges still contain microorganisms, some of which may be pathogens, and are unlikely to be odour-free. Depending on the choice of disposal route it may thus be desirable to stabilize the sludge.

Thermal drying

During the past few years there has been a growth of interest in sludge dryers for further dewatering and stabilization of sludge cake from mechanical dewatering processes. The technology employed has been adapted from the process dryers which are widely used in many industries and involve direct or indirect contact between the sludge particles and a heat source. This occurs either in a rotating drum or in a fluidized bed configuration and results in an odour-free, stable material with a solids content of 70–90 per cent which can be pelletized to provide a controlled quality product. This dried sludge is at least partly pasteurized and can be bagged and sold as a soil conditioner. Thermal drying may also be used as pretreatment before incineration. The vapour and emissions from a sludge dryer must be handled effectively to prevent air pollution problems. This can be achieved by recycling all gaseous emissions into the burners of the dryer under controlled conditions.

Chemical stabilization

Lime has been used in sludge conditioning prior to mechanical dewatering for many years but over the past ten years the concept of further treatment of sludge cake with lime or other chemicals has received some attention and several proprietary processes have been developed. The addition of lime to cake raises the pH to around 12 and causes heating and the release of ammonia which kills most microorganisms in the sludge. Proprietary processes add cement kiln dust or power station fly ash to sludge cake which again raises the pH to around 10–12 and generates heat to produce a stable product with very low levels of microbial activity.

Composting

Composting is an aerobic biological oxidation process for the stabilization of organic matter which is well known to keen gardeners but which is also used to some extent for the stabilization of wastewater sludges. In composting, aerobic microorganisms convert much of the organic matter into carbon dioxide leaving a relatively stable odour-free substance which has some value as a soil conditioner. The composting process involves careful control of moisture content which must be between 40 and 60 per cent and liquid or cake sludge must be mixed with a bulking agent such as sawdust, shredded paper or domestic refuse.

The simplest method of composting involves the use of piles of the mixed sludge and bulking agent which are turned regularly by a front-end loader or similar machine for a period of several weeks. This ensures that aerobic conditions are maintained and that a homogeneous final compost is produced.

Other composting techniques include static piles which are not mixed but which are aerated by the introduction of air through a distribution system at the base, and composting reactors which use vertical or inclined kiln type vessels.

Since there is only a small reduction in volume during composting the process is usually only attractive if a reliable outlet for the product exists in the form of nearby agricultural or horticultural activities.

18.5 Sludge disposal

Current wastewater sludge production in the UK amounts to around 1.2 megatonnes of dry solids each year and for the EU as a whole there are about 6.5 megatonnes of dry solids produced annually. These sludge production values will be significantly increased, possibly by as much as 50 per cent, as the urban wastewater treatment directive is implemented over the period up to 2005. Numerous new wastewater treatment plants will have to be constructed in countries where the whole population is not currently served, discharges to the sea will require treatment and nutrient removal in sensitive areas will produce additional sludge. Table 18.1 shows the disposal routes for sewage sludge in the UK before and after the implementation of the prohibition of sea dispersal.

Table 18.1 Sewage sludge disposal routes in the UK

Disposal route	1996 (%)	2005 (%, estimated)
Agricultural land	49	58
Sea dispersal	23	0
Landfill	10	10
Incineration	7	25
Other	11	7

Sludges from water treatment operations in the UK amount to around 100 kilotonnes per year and this smaller amount, together with the relatively inert nature of waterworks sludges, means that they pose little problems in relation to the wastewater sludge production. Waterworks sludges are usually either landfilled or in some cases sprayed on moorland or forestry plantations. Softening sludges with a high calcium carbonate content can be utilized for agricultural purposes. Recovery of aluminium sulphate from alum sludges is technically possible but uneconomic.

Disposal to agricultural land

Sludge from the treatment of domestic sewage has some value as a soil conditioner since it contains significant amounts of nitrogen and phosphorus. When industrial wastewaters are treated at a municipal sewage works the possibility of the accumulation of undesirable constituents, such as heavy metals, in the sludge must be guarded against if the sludge is to be used agriculturally. In some situations the acceptability of industrial wastewaters into municipal sewers may be governed by the constraints imposed because of the ultimate disposal route for the sludge. Raw sewage sludges, although higher in nutrient content than digested sludges, are likely to have objectionable odours and contain large numbers of potentially pathogenic microorganisms. Most countries now require anaerobic digestion of sewage sludges before any application to agricultural land unless the sludge is injected directly below the soil surface. There are growing pressures for the adoption of further restrictions on the use of sewage sludge on agricultural land to reduce any health risks. Application to land can be via direct sub-surface injection or surface spraying of liquid sludges. In situations or climates where additional water on the land is undesirable it may be necessary for the sludge to be dewatered before application. Alternatively, holding facilities will have to be provided to store liquid sludge until it can be disposed of to land.

As well as the potential health hazards from microorganisms in wastewater sludges there may also be hazards from the presence of heavy metals which can become concentrated in food chains. Thus if sludge applications are not carefully controlled there are situations in which the metal contents of a sludge can result in possibly harmful accumulations in crops grown on the sludge-treated soil. EC regulations (Sewage Sludge Directive 86/278) place a statutory requirement on sludge-producing authorities to monitor sludge for zinc, copper, nickel, cadmium, lead, mercury and chromium. This information, together with soil analyses, enables sludge loadings to be controlled so that limits for these metals are not exceeded. Care must also be taken to ensure that significant amounts of synthetic organic chemicals are not present in sludges applied to agricultural land. Depending upon agricultural practices, soil type, climate and sludge composition a land disposal area of $20\,m^2$/person upwards is likely to be necessary. Tables 18.2 and 18.3 illustrate the requirements of the UK Department of the Environment for agricultural use of sewage sludges which implement the EC regulations but which impose additional requirements on certain metals.

Other sludge disposal routes

Landfilling

Considerable amounts of wastewater sludges are dumped in old quarries or used as landfill material. Such disposal methods can be of value in land reclamation projects but the possibility of groundwater contamination must always be

Table 18.2 Examples of effective sludge treatment processes for application to agricultural land (Department of the Environment, 1989)

Process	Description
Pasteurization	Minimum of 30 min at 70°C or 4 h at 55°C followed by mesophilic anaerobic digestion
Mesophilic anaerobic digestion	Mean retention of at least 12 days in temperature range of 35°C ±3°C or at least 20 days at 25°C ±3°C followed by secondary retention of at least 14 days
Thermophilic aerobic digestion	Mean retention period of at least 7 days with a minimum of 55°C for at least 4 h
Composting (windrows or aerated piles)	Temperature must be maintained at 40°C for at least 5 days and for 4 h a minimum of 55°C must be reached in the body of the pile
Lime stabilization (liquid sludge)	Addition of lime to raise pH to greater than 12 and sufficient to ensure that pH is not less than 12 for a minimum period of 2 h
Liquid storage	Minimum storage period of three months for untreated sludge
Dewatering and storage	Conditioned, untreated dewatered sludge – minimum storage three months Anaerobically treated before dewatering – minimum storage 14 days

Table 18.3 Maximum permissible concentrations of potentially toxic elements in soil after application of sewage sludge and maximum annual rates of addition (Department of the Environment, 1989)

Potentially toxic element (PTE)	Maximum permissible concentration of PTE in soil (mg/kg dry solids)				Maximum permissible average annual rate of PTE addition over a 10-year period (kg/ha)
	pH 5.0–5.5	pH 5.5–6.0	pH 6.0–7.0	pH > 7.0	
Zinc	200	250	300	450	15
Copper	80	100	135	200	7.5
Nickel	50	60	75	110	3
	For pH 5.0 and above				
Cadmium	3				0.15
Lead	300				15
Mercury	1				0.1
Chromium	400 (provisional)				15 (provisional)
Molybdenum	4				0.2
Selenium	3				0.15
Arsenic	50				0.7
Fluoride	500				20

considered and it is becoming increasingly difficult to find suitable sites where environmental regulatory bodies will permit disposal of sewage sludges.

Sea dispersal

In coastal communities, or in those with easy access to the sea, disposal of sludge at sea has long been a favoured practice. Properly controlled disposal of sewage sludges in suitable deep water locations has not been shown to produce any significant detrimental environmental effects even when the practice has been carried out over many years. Nevertheless, environmental pressure groups have succeeded in persuading legislators in many parts of the world to ban sludge disposal at sea. An EC directive prohibits the disposal of sewage sludge to seas around Europe after 1998. Implementation of this directive will require the use of costly alternatives which may in the end have environmental effects which are more detrimental than the current sea disposal practices.

Incineration

Transport costs for sludge disposal become significant as the distance to the disposal site increases and in some locations the cost of transport may rule out all the disposal routes described above, leaving incineration as the only feasible technique. If sludges contain toxic substances incineration may again be the only environmentally acceptable route. Early sludge incinerators were often of the rotary multiple-hearth type and most were somewhat unsatisfactory in operation, with high maintenance costs and poor-quality stack emissions which became unacceptable under IPC constraints. More recently, fluidized bed furnaces have become popular and several new installations have been completed in the UK. Sewage sludges if dewatered to 30–35 per cent solids content usually have a sufficiently high calorific value (16–20 MJ/kg dry solids) to support self-sustaining combustion although many incinerators use a small amount of fuel oil or gas for optimum performance. After incineration at about 850°C the residual ash amounts to 5–10 per cent of the original solids and the inert material can usually be disposed of fairly easily as fly ash, although the heavy metal content of the ash may cause problems in finding an acceptable disposal site.

Emission rates from the incineration of sewage sludge are shown in Table 18.4

Early sludge incinerators operated under somewhat undemanding controls on the quality of stack emissions and were usually provided with a simple ash quenching system and a gas scrubber to remove particulates. Much stricter stack emission standards now apply to sewage sludge incinerators as indicated in Table 18.5.

Table 18.4 Typical emissions rates from sewage sludge incineration

Constituent	Emission rate (kg/tonne dry solids)
Particulates	16
NO_x	2.5
SO_x	0.5
Hydrocarbons	0.5

Table 18.5 Stack emission standards for sewage sludge incinerators

Constituent	MAC (mg/m³ at STP)	
	BimSchV (Germany)	EA (UK)
Dust	10	20
Hydrogen chloride	10	30
Hydrogen fluoride	1	2
Sulphur dioxide	50	300
Carbon monoxide	50	50
Nitrogen oxides	200	
TOC	10	
Cadmium/thallium	0.05	0.05
Mercury	0.05	0.05
Other heavy metals	0.05	0.5
Dioxins	0.1 ng/m³	0.1 ng/m³

Achievement of these stack emission standards requires an effective gas cleaning stage usually involving heat recovery, electrostatic precipitation and wet alkali scrubbing with provision for treatment and disposal of the scrubber liquor.

Vitrification or sludge melting

Some success has been achieved in Japan with high-temperature vitrification of sludge cake or ash from incineration to produce glazed floor tiles, road sub-base aggregate and similar materials. The process is complex and costly involving heating to 1200–1500°C at which temperatures the organic matter burns off and the inorganic constituents liquefy. On cooling, a slag or glass is formed which is very hard and which appears to bind the heavy metals in a form which prevents them leaching out of the material. The high temperatures employed in the process mean that the stack emissions are free from complex hydrocarbons and other undesirable constituents. The final volume of solids produced in the vitrification process is about half that produced by conventional incineration of the same sludge and this reduction in volume for ultimate disposal is the main attraction in Japan where land prices are very high.

Pyrolysis

Some work in Canada and Australia has been undertaken on a process which heats undigested sludge to a temperature of about 450°C in the absence of oxygen. Under these conditions some of the sludge vaporizes and the vapour can be converted catalytically to a hydrocarbon with similar properties to diesel oil. In demonstration plants the yield of oil is 200–300 l/tonne dry solids. More development work is needed before the viability of large-scale application of the process can be determined and the economics of oil production from sludge would be controlled by market prices for oil.

Choice of disposal route

Selection of a disposal route for wastewater sludges must be made on the basis of a critical assessment of conditions and constraints affecting a specific plant. Use of a disposal route which is cost effective and which provides for beneficial use of the sludge is favoured and thus in many situations disposal of liquid sludge to agricultural land would usually be the first choice, followed closely by application to dedicated land owned by the organization producing the sludge. More complex routes only become appropriate when the nature of the sludge prevents land application or where suitable land is not available within a reasonable distance from the wastewater treatment plant. Table 18.6 gives some

Table 18.6   Relative costs of some sewage sludge disposal routes

Disposal route	Relative capital cost	Relative operating cost
Digestion and disposal to agricultural land	1.8	2
Digestion and disposal to dedicated land	1.8	1
Dewatered and tipped	1	2.5
Dewatered and dried	2	2.7
Dewatered and incinerated	3.8	3

relative costs for various disposal routes. Typical UK capital costs for incineration, which is normally the most expensive solution, are £450–800/tonne dry solids for capacities of 250 to 25 tonnes dry solids per day, respectively. Incineration operating costs are usually in the range £50–120/tonne dry solids.

References

Baskerville, R. C. and Gale, R. S. (1968). A simple automatic instrument for determining the filtrability of sewage sludges. *Wat. Pollut. Control*, **67**, 233.

Coackley, P. (1955). Research on sewage sludges carried out in the Civil Engineering Department of University College, London. *J. Proc. Inst. Sew. Purif.*, 59.

Department of the Environment (1989). *Code of Practice for Agricultural Use of Sewage Sludge*. London: HMSO.

Further reading

American Water Works Association (1989). *Sludge, Handling and Disposal*. Denver: AWWA.

Bruce, A. M. (1995). *Sewage Sludge: Utilization and Disposal.* London: CIWEM.

Bruce, A. M., Pike, E. B. and Fisher, W. J. (1990). A review of treatment process options to meet the EC Sludge Directive. *J. Instn Wat. Envir. Managt*, **4**, 1.

Carberry, J. B. and Englande, A. J. (eds) (1983). *Sludge Characteristics and Behavior.* The Hague: Nijhoff.

Carroll, B. A., Caunt, P. and Cunliffe, G. (1993). Composting sewage sludge: Basic principles and opportunities in the UK. *J. Instn Wat. Envir. Managt*, **7**, 175.

Davis, R. D. (1989). Agricultural utilization of sewage sludge: a review. *J. Instn Wat. Envir. Managt*, **3**, 351.

Frost, R. C. (1988). Developments in sewage sludge incineration. *J. Instn Wat. Envir. Managt*, **2**, 465.

Geertsema, W. S., Knocke, W. R., Novak, J. T. and Dove, D. (1994). Long term effects of sludge application to land. *J. Amer. Wat. Wks Assn*, **86**(11), 64.

Hall, J. E. (1995). Sewage sludge production, treatment and disposal in the EU. *J. C. Instn Wat. Envir. Managt*, **9**, 335.

Hoyland, G., Dee, A. and Day, M. (1989). Optimum design of sewage sludge consolidation tanks. *J. Instn Wat. Envir. Managt*, **3**, 505.

Hudson, J. A. and Lowe, P. (1996). Current technologies for sludge treatment and disposal. *J. C. Instn Wat. Envir. Managt*, **10**, 436.

Kane, M. J. (1987). Coventry area sewage sludge disposal scheme: development of strategy and early operating experiences. *J. Instn Wat. Envir. Managt*, **1**, 305.

Lowe, P. (1988). Incineration of sewage sludge – a reappraisal. *J. Instn Wat. Envir. Managt*, **2**, 416.

Lowe, P. (1995). Developments in the thermal drying of sewage sludge. *J. C. Instn Wat. Envir. Managt*, **9**, 306.

Matthews, P. (1992). Sewage sludge disposal in the UK: A new challenge for the next twenty years. *J. Instn Wat. Envir. Managt*, **6**, 551.

Polprasert, C. (1996). *Organic Waste Recycling* (2nd edn). Chichester: John Wiley.

Tebbutt, T. H. Y. (1992). Japanese Sewage Sludge Treatment, Utilization and Disposal. *J. Instn Wat. Envir. Managt*, **6**, 628.

Tebbutt, T. H. Y. (1995). Incineration of wastewater sludges. *Proc. Instn Civ. Engnrs Wat., Marit. & Energy*, **112**, 39.

Vesilind, P. A. (1979). *Sludge and its Disposal.* Ann Arbor: Ann Arbor Science Publishers.

Problems

1. A sewage flow of $1\,m^3/s$ containing $450\,mg/l\,SS$ is given primary sedimentation to remove 50 per cent of the SS before discharge to the sea. Calculate the daily volume of sludge produced if it is drawn off from the sedimentation tanks at 4 per cent solids. Assume solids have relative density 1.4. ($480\,m^3$)

2. Given a volume of $100\,m^3$ of sludge at 95 per cent moisture content determine the volume occupied by the sludge when dewatered to 70 per cent moisture. Assume solids have relative density 1.4. ($15.5\,m^3$)

3. The following results were obtained from a specific resistance to filtration determination on a sample of activated sludge:

Time, t (s)	Volume of filtrate, V (ml)
0	0.0
60	1.4
120	2.4
240	4.2
480	6.9
900	10.4

Vacuum pressure = 97.5 kPa
Filtrate viscosity = 1.001×10^{-3} N s/m^2
Solids content = 7.5% (75 kg/m^3)
Area of filter = 4.42×10^{-3} m^2

Plot the values of t/V against V and hence obtain the slope and calculate the specific resistance. (2.4×10^{14} m/kg)

19

Tertiary treatment, water reclamation and re-use

As environmental pressures increase and the finite limitations of water resources become apparent it is not surprising that in addition to conventional water and wastewater treatment processes attention has become focused on additional or alternative processes to achieve new objectives.

Although conventional sewage treatment processes can achieve a considerable degree of purification, this may be insufficient in situations with little dilution or where potable water abstractions or water-based recreational activities occur downstream. In such cases an additional stage of treatment to remove most of the remaining SS and the associated BOD is often stipulated by the regulatory authorities. This type of additional removal is usually termed tertiary treatment. The increasing concern at the accelerated eutrophication of some surfacewaters has resulted in the development of nutrient removal techniques and the EC urban wastewater treatment directive requires nutrient removal for discharges in sensitive areas.

In areas where water resources are limited, it may become necessary to utilize wastewater effluents, brackish groundwaters or even seawater to satisfy the demands of domestic and industrial consumers. Many domestic uses of water such as toilet flushing do not need potable water and recycled effluent may be appropriate for a dual supply. Some industrial uses of water can be satisfied quite easily by conventionally or tertiary treated sewage effluent and such sources can be perfectly acceptable for irrigation use under carefully controlled conditions. For more demanding uses it may be necessary to utilize treatment processes specifically designed to remove particular impurities which are not affected by conventional treatment processes. The conversion of saline waters into potable supplies requires the removal of dissolved inorganic constituents that are unaffected by conventional water treatment processes.

19.1 Tertiary treatment

Although a conventional sewage-treatment plant incorporating primary settlement, biological oxidation and final settlement may be able to produce an effluent of better than 30 mg/l SS and 20 mg/l BOD at times, the reliable production of an effluent significantly better than 30:20 requires some form of tertiary treatment. The Royal Commission recommendation that 4 mg/l BOD be the limiting level of pollution in receiving waters to prevent undesirable conditions has been found to be rather unrealistic since there are many rivers in the UK which have BOD levels in excess of 4 mg/l but which have high DO levels and are used for water-supply purposes. The need for tertiary treatment in a particular situation should therefore be assessed in the light of the circumstances relevant to that situation, i.e. dilution, reaeration characteristics, downstream water use and water quality objectives for the receiving water.

The main reason for limiting SS in effluents is that they may settle on the stream bed and inhibit certain forms of aquatic life. Flood flows may resuspend these bottom deposits and exert sudden oxygen demands. Settlement does not, however, always occur in waters which are naturally turbid. Effluent SS levels may not be of great significance in themselves, although they do of course influence the BOD of the effluent. In some cases where a restriction on SS levels is desirable a reduction in BOD to below 20 mg/l may also be necessary. This is not always so, however, and the need for BOD reduction should be examined separately. The BOD determination is by its very nature a somewhat unreliable parameter and BOD standards should not be specified with excessive accuracy. Possible BOD standards could be 20, 15, 10 and exceptionally 5 mg/l.

Most forms of tertiary treatment used in the UK have been aimed at removal of some of the excess SS in the effluent from a well operated conventional works. Tertiary treatment should be considered as a technique for improving the quality of a good effluent and not as a method of trying to convert a poor effluent into a very good-quality discharge. Removal of SS from an effluent gives an associated removal of BOD due to the BOD exerted by the suspended matter. There is a good deal of evidence to show that for normal sewage effluents the removal of 10 mg/l SS is likely to remove about 3 mg/l BOD.

Various methods of tertiary treatment are available, several of which are based on processes used in water treatment.

Rapid filtration

This process is frequently employed in large plants. Most installations are based on the downflow sand filter which has been used in water-treatment plants for many years. More efficient forms of filter, including mixed media beds and upflow units, have been used with some success, but in many cases the downflow unit is adopted because of its simplicity and reliability. The variable nature of the SS present in the effluent from final settling tanks makes prediction of the performance of any tertiary treatment unit difficult. Because of the wide variation in filtration characteristics of suspended matter it is always advisable to carry out experimental work on a particular effluent before proceeding with design work.

It is generally assumed that rapid gravity filters operated at a hydraulic loading of about 200 m³/m² day should remove 65–80 per cent SS and 20–35 per cent

BOD from a 30:20 standard effluent. Because of the relatively short time which elapses between backwashes (24–48 h generally), little biological activity occurs and thus rapid filters are not likely to achieve any significant oxidation of ammonia. The SS removal is not significantly affected by the hydraulic loading within quite wide variations and there is little benefit in using sand smaller than 1.0–2.0 mm grading.

Slow filtration

On small works slow sand filters are sometimes employed for tertiary treatment at loadings of 2–5 m^3/m^2 day. Slow filters have low operation and maintenance costs, but their relatively large area requirements normally rule them out for other than small installations. They can be expected to remove 60–80 per cent SS and 30–50 per cent BOD. Slow filters provide a significant amount of biological activity, thus encouraging BOD removal and providing a degree of nitrification. In addition they can provide significant removals of bacteria and other microorganisms.

Microstraining

Microstrainers have been utilized for tertiary treatment since 1948 and a number of installations are in operation. They have the advantage of small size and can thus easily be placed under cover. Removals of SS and BOD depend upon the mesh size of the fabric used and the filtrability characteristics of the suspended matter. Reported removals range from 35 to 75 per cent SS and 12 to 50 per cent BOD. Microstraining should reliably give an effluent of 15 mg/l SS, and 10 mg/l SS should be possible with a good final tank effluent. Typical filtration rates are 400–600 m^3/m^2 day. Biological growths on the fabric which could cause clogging and excessive head loss can normally be controlled by UV lamps.

Upward-flow clarifier

This technique was originally developed as a means of obtaining better quality effluents from conventional humus tanks on small bacteria bed installations. The process is based on passing the tank effluent through a 150 mm layer of 5–10 mm gravel supported on a perforated or wedge-wire plate near the top of a horizontal flow humus tank with surface overflow rates of 15–25 m^3/m^2 day. Passage through the gravel bed promotes flocculation of the suspended matter and the floc settles on top of the gravel. Accumulated solids are removed intermittently by drawing down the liquid level below the gravel bed. Removals of 30–50 per cent SS can be achieved, dependent upon the size of gravel and the type of solids. Similar results have been achieved with wedge wire in place of gravel and there is evidence that many types of porous materials can be used to promote flocculation.

Grass plots

Land irrigation on grass plots can provide a very effective form of tertiary treatment particularly suitable for small communities. Effluent is distributed over grassland, ideally with a slope of about 1 in 60, and collected in channels at the bottom of the plot. Hydraulic loadings should be in the range $0.05-0.3\,m^3/m^2$ day and 60–90 per cent SS and up to 70 per cent BOD removals can be achieved. Short grasses are preferable, but the sowing of special grass mixtures does not appear to be worthwhile. The area should be divided into a number of plots to permit access, for grass and weed cutting, since growth is likely to be prolific due to the nutrients present in effluents.

Reed beds

As discussed in Chapter 14, the use of reed beds for tertiary treatment on small wastewater treatment plants is becoming common and they are capable of producing good quality effluents under the right conditions. At loadings of $1\,m^2$/person reed beds after RBC or conventional filtration should reliably produce effluents with average SS and BOD levels of around 5 mg/l and with significant removals of ammonia.

Lagoons

Storage of effluent in lagoons or maturation ponds provides a combination of sedimentation and biological oxidation depending upon the retention time. With short retention times (2–3 days) the purification effect is mainly due to flocculation and sedimentation, with SS removals of 30–40 per cent being likely.

With longer retention times (14–21 days) the improvement in quality may be very marked with 75–90 per cent SS, 50–60 per cent BOD and 99 per cent coliform removals. Heavy algal growths may, however, occur with these larger ponds, and at times the escape of algae from the pond can result in relatively high SS and BOD levels in the final discharge. The improvement in bacteriological quality in lagoons is greater than provided by most other forms of tertiary treatment, with the possible exception of grass plots and reed beds, and is of particular interest if the receiving water is a raw water source. In UK conditions it would appear that a retention time of about eight days provides the most satisfactory overall performance, since longer retention times are more likely to give rise to excessive algal growths.

19.2 Water reclamation and re-use

Increasing demands for water will in the future require the development of new sources, some of which will contain water of a quality inferior to that judged acceptable in the past for water-supply purposes. In densely populated areas, much of the increased demand may have to be satisfied by abstractions from lowland rivers which are likely to contain significant proportions of sewage effluent and industrial wastewaters.

The direct re-use of sewage effluent to satisfy a number of industrial water requirements is already an accepted practice, with consequent economies in cost and one which also serves to release supplies of better grade water which would otherwise be used industrially. The use of sewage as a source of potable water is technically feasible at the present time, but would be relatively costly and its use would probably produce psychological objections from the consumers. Such direct re-use would require the adoption of additional processes mainly physico-chemical in nature, which are likely to be fairly costly. The adoption of such techniques should therefore only be as the result of careful cost–benefit analysis of the situation.

In this context it is important to differentiate between different forms of wastewater re-use, being

- indirect non-potable, in which treated wastewater effluents are discharged to watercourses or aquifers used for agricultural and/or industrial abstractions
- direct non-potable, in which treated wastewater effluents are conveyed directly to an agricultural, industrial or landscape point of use
- indirect potable, in which treated wastewater effluents are discharged to water courses or aquifers used as sources of potable water
- direct potable, in which treated wastewater effluents are connected directly to water supply systems.

The first three forms of wastewater re-use are already widely employed but direct potable re-use is rarely used on a routine basis.

Conventional water treatment (coagulation, sedimentation, filtration and disinfection) was originally developed for the removal of suspended and colloidal solids from raw waters together with limited removal of the soluble organics responsible for the natural colour in water from upland catchments. Certain soluble constituents, such as those responsible for hardness, can be removed by the incorporation of precipitation or ion-exchange processes. Using such techniques it is possible to produce an acceptable water from relatively heavily polluted sources, but it must be appreciated that there are limits to the levels of certain types of impurity which can be satisfactorily dealt with by conventional water treatment. Indeed some impurities may be completely unaltered by normal water-treatment methods. Nevertheless, there are a number of examples of situations where emergency conditions have required the treatment of heavily polluted waters without danger to the consumer. In such circumstances it may be necessary to distinguish between potable and wholesome supplies, since it is by no means certain that a potable supply would also be considered wholesome.

Temporary war-time use of the River Avon at Ryton (Pugh, 1945) is an illustration of the results (Table 19.1) of subjecting a raw water from a river

Table 19.1. *River Avon supply 1944 (Pugh, 1945)*

Characteristic (mg/l except where noted)	Raw water		Treated water	
	Average	*Range*	*Average*	*Range*
Turbidity (units)	25	5–425	0.6	0–7.6
Colour (°H)	21	12–70	4	0–16
TDS	768	516–1012	759	532–968
Cl	48	29–68	50	33–69
Ammonia nitrogen	1.4	0.6–17.4	1.3	0–17.4
Nitrate nitrogen	7	2–10	8	3–13
E. *coli*/100 ml	9 700	200–25 000	0	0–0
Bacterial colonies/ml (37°C)	5 100	20–29 000	4	0–20
Bacterial colonies/ml (22°C)	87 000	2400–600 000	11	1–60

heavily polluted by sewage to conventional water treatment. The data illustrate the limitation of conventional treatment as regards the removal of dissolved solids, chloride, nitrogen compounds and dissolved organics. Later stages of the abstraction in Ryton involved the addition of a high-rate bacteria bed ahead of the main plant. This nitrified the ammonia, which reduced disinfection problems and was also useful in oxidizing organic matter responsible for taste and odour troubles.

A more direct form of effluent re-use took place for a short period in 1956–7 at Chanute in the USA (Metzler *et al.*, 1958) where drought conditions caused a failure of the water supply and sewage effluent was recycled into the water-treatment plant. Both treatment plants were of conventional design and at the end of a five-month period, when it was estimated that about ten cycles had taken place, it was felt that the limit had been reached. Table 19.2 gives an indication of the build-up of impurities during the operation and again illustrates the inability of conventional treatment processes to deal with dissolved solids, nitrogen compounds and most dissolved organics.

Table 19.2 Re-use at Chanute (Metzler *et al.*, 1958)

Characteristic (mg/l)	Original water	After ten cycles
TDS	305	1139
Cl	63	520
Ammonia nitrogen	–	10
Nitrate nitrogen	1.9	2.7
COD	–	43
ABS	–	4.4
SO$_4$	101	89
PO$_4$	–	3.9

In the 1960s and 70s full-scale studies (van Vuuren *et al.*, 1980) were undertaken at Windhoek to assess the viability of using reclaimed wastewater effluent to augment the available freshwater supplies. The effluent from a conventional biological filtration plant was denitrified and passed through maturation ponds before entering the reclamation stages which included alum coagulation and flotation to remove algae from the maturation ponds, liming, aeration, sedimentation, rapid filtration, activated carbon adsorption and chlorination. The intention was to blend the reclaimed water with freshwater in the ratio of 1:2 and in this way reach a stable composition.

Thus although it is possible to obtain a marked improvement in the quality of sewage effluent by the application of conventional water treatment processes, it seems likely that the product water would be unacceptable as a potable supply on a regular basis. The inevitable presence of trace organics and DBPs in the water would make it impossible to meet WHO drinking water guideline values.

In parts of California and in certain other dry areas of the USA considerable work has been undertaken to evaluate the potential of water reclamation from wastewater effluents. In Denver a demonstration plant has been constructed (Lauer *et al.*, 1985) to convert secondary sewage effluent into water of potable quality. The complex process chains used have included lime clarification, recarbonation, filtration, selective ion exchange, first-stage carbon adsorption, ozonization, UV disinfection, second-stage carbon adsorption, reverse osmosis, air stripping and chlorine dioxide disinfection. The plant also includes a selective ion exchange regeneration and recovery system and a fluidized bed carbon regeneration furnace. A five-year operational programme was undertaken to evaluate the viability and acceptability of the process but although water of potable quality was produced none was in fact used for public consumption. Regardless of the quality of such reclaimed water, it is doubtful whether it would readily gain consumer acceptance.

In a number of countries in the Middle East there are installations which take the final effluent from conventional wastewater treatment plants and by the use of process chains such as ozonization, coagulation and sedimentation, filtration, chemical stabilization, membranes and chlorination can produce potable quality water. In most, if not all, cases use of the product water is restricted to secondary domestic purposes and landscape irrigation. In many Eastern Mediterranean countries reclaimed wastewater is a major source of irrigation water for high-value cash crops. In Israel, for example, reclamation schemes employing conventional wastewater treatment processes followed by stabilization ponds and recharge to provide soil–aquifer treatment are widely used. By 2010 it is estimated that reclaimed wastewater will provide 20 per cent of the total water supply in Israel and 33 per cent of irrigation demand.

In Japan the dense urban development in many cities has placed severe demands on water resources and encouraged an active programme of effluent re-use. In Tokyo filtered and disinfected wastewater effluent is used for toilet flushing in a high-rise development area, in landscape water features and for industrial process water.

For such uses as landscape irrigation, fodder irrigation, groundwater recharge and certain industrial processes, reclaimed wastewater is widely used in water-short parts of the world and water quality standards have been recommended by WHO (Table 19.3) and USEPA (Table 19.4). It is worth noting that simple chemical coagulation of secondary sewage effluent followed by disinfection can produce a water which is suitable for many non-potable uses. The potential health hazards inherent in the use of sewage effluent must always be guarded against in any re-use situation. Sensible controls on applications can reduce hazards from reclaimed sewage effluents to acceptable levels.

Table 19.3 WHO recommended microbiological quality guidelines for wastewater use in agriculture (World Health Organization, 1989)

Category	Re-use conditions	Exposed group	Intestinal nematodes (eggs/litre)	Faecal coliforms (cells/litre)	Treatment to achieve the microbiological quality
A	Irrigation of crops likely to be eaten uncooked, sports fields, public parks	Workers, consumers, public	<1	<1000	Series of stabilization ponds
B	Irrigation of cereal crops, industrial and fodder crops, pasture, trees	Workers	<1	No std	8–10 day retention in stabilization ponds
C	Localized irrigation of crops in category B if no human exposure	None	N/a	N/a	At least primary sedimentation

Table 19.4. USEPA typical guidelines for effluent re-use (United States Environmental Protection Agency, 1992)

Type of re-use	Reclaimed quality

Urban re-use
  Landscape irrigation
  Toilet flushing
  Recreational lakes
	pH 6–9, BOD < 10 mg/l, turbidity < 2 NTU; no faecal coliforms/100 ml; 1 mg/l residual chlorine

Agricultural re-use
  Food crops, commercially processed
  Surface irrigation, orchards, vineyards
  Non-food crops
	pH 6–9, BOD < 30 mg/l, SS < 30 mg/l; faecal coliforms < 200/100 ml; 1 mg/l residual chlorine

Groundwater recharge
  Potable aquifers
  Indirect re-use
	pH 6.5–8.5, turbidity < 2 NTU; no faecal coliforms/100 ml; 1 mg/l residual chlorine; other parameters as potable standards

It will be clear from the foregoing comments that to obtain a wholesome potable supply of water from heavily polluted sources such as sewage effluent, conventional water treatment alone is not sufficient. To achieve the desired end quality a number of alternative courses of action could be adopted, as listed below.

1. Provide additional treatment stages at wastewater and/or water treatment plants to deal with contaminants not affected by normal treatment.
2. Provide a completely new form of wastewater and/or water treatment.
3. Use conventional treatment processes and blend the finished water with another water of higher quality so that the mixture is of acceptable quality.
4. Dispense with separate wastewater and water-treatment facilities (and the intervening receiving water) and introduce an integrated water reclamation facility.

The second and fourth possible actions are certainly feasible from a technological point of view at the present time. The use of distillation or reverse osmosis would permit the production of an acceptable water at a cost similar to that of producing freshwater from seawater. Such techniques could permit the direct recycling of sewage effluent, although a certain amount of make-up water would be necessary because of losses in the system. Completely closed-cycle systems for water and wastewater are of considerable interest in space-vehicle development, but in this application costs are of secondary importance. It would thus seem likely that for large-scale use of wastewater as a raw water source the first and third courses of action are most likely to have the greatest application.

The continuance of the system by which wastewater is treated and discharged to a receiving water from which abstractions may be made for water-supply purposes has much to commend it. The presence of dilution water is useful (assuming that the receiving water is of better quality than the effluent) and the concentration of non-conservative pollutants will be reduced by self-purification between discharge and abstraction points. Consumer acceptance of such an indirect recycling scheme is also likely to be better than for direct recycling. In this context it is necessary to consider the characteristics of conventionally treated sewage effluent which would be undesirable in water treatment terms. Table 19.5 shows typical analyses of crude sewage and sewage effluents compared with EC recommendations for raw water quality. It is clear that a number of the characteristics of sewage effluent are likely to cause problems as regards treatment in a conventional water works. In particular, the high contents of non-biodegradable organics, total solids, ammonia and nitrate nitrogen are likely to be troublesome. In addition, the very high levels of bacteriological (and viral) impurity would be such as to cause concern to water-treatment authorities.

Dissolved organics, which are largely non-biodegradable or, at least, only slowly broken down by biological means, are found in sewage effluents in the form of many different compounds whose presence is indicated by high TOC and COD values with relatively low BOD values. Some of the organics in sewage

Table 19.5. Typical sewage characteristics and water quality limits

Characteristic	Crude sewage	Conventional effluent (30:20 standard)	Sand-filtered effluent	EC raw water directive for A3 treatment (see Table 2.4)	
				Guide limit	Mandatory limit
BOD (mg/l)	400	20	10	7	
COD (mg/l)	800	100	70	30	
BOD/COD ratio	0.50	0.20	0.14	0.23	
TOC (mg/l)	300	35	25		
SS (mg/l)	500	30	10		
Cl (mg/l)	100*	100*	100*	200	
Organic N (mg/l)	25	0	0	3	
Ammonia N (mg/l)	25	5†	2†	2	4
Nitrate N (mg/l)	0	20†	24†		11.3
Total solids (mg/l)	1000*	1000*	1000*	(1000 μS/cm)	
Total hardness (mg/l)	250*	250*	250 *		
PO$_4$ (mg/l)	10	6	6	0.7	
ABS (mg/l)	2	0.2	0.2		0.5
Colour (°H)		50	30	50	200
Coliforms/100 ml	10^7	10^6	10^5	5×10^4	

* Concentration depends to some extent on quality of carriage water.
† Concentration depends upon degree of nitrification achieved in wastewater treatment.

effluents are believed to be similar to the compounds which produce natural colour in upland catchments. It is, however, certain that other organic compounds may be more troublesome in water, particularly as regards the formation of tastes and odours. They are also likely to increase the chlorine demand and the possibility of toxic effects must also be considered.

Total solids is a somewhat vague parameter for the assessment of water quality, since it gives no indication of the source or nature of the impurities. Clearly, a few milligrams per litre of certain compounds could be toxic, whereas several hundred of other compounds would be quite harmless. There is unfortunately a great lack of information about the effects on humans of inorganic compounds in water. In situations where a water is within normal potable standards with the exception of total solids there is often a willingness to accept a total solids level higher than that considered desirable by WHO.

Nitrogen compounds in the form of ammonia or nitrate are present in all sewage effluents and are undesirable in potable water because of disinfection problems and promotion of biological growths with resulting tastes and odours (ammonia) and due to the potential health hazard to young babies from nitrates.

Phosphates, another normal constituent of sewage effluent, may also be troublesome in water-treatment processes because of the inhibiting effect they can sometimes produce on coagulation reactions.

The existence of large numbers of microorganisms in a raw water is always of great concern to water-treatment authorities, but since disinfection is a well-established process there seems no reason why bacteriologically satisfactory water could not be produced from sewage effluent. The inactivation of viruses is, however, somewhat less predictable and thus their presence in raw water is particularly undesirable. The possible existence of protozoan cysts, which are less readily inactivated by normal disinfection, also needs consideration in re-use situations.

19.3 Removal of impurities not amenable to conventional treatment

Dissolved organics

Biological oxidation of polluted raw water on a high-rate bacteria bed can achieve some reduction in BOD, possibly of about 20 per cent. Such installations are, however, usually installed primarily for the oxidation of ammonia. More substantial removal of soluble organics can be achieved by adsorption on activated carbon. Both powdered and granular forms of activated carbon can be employed to give relatively high removals of COD and TOC, although a residual of unadsorbable material may remain. For intermittent use, powdered carbon may be satisfactory with the addition being made to the coagulation/sedimentation stage or to the filters. Where continuous use of activated carbon is necessary the granular form is more appropriate and provision must be made for regeneration either on-site or in a central facility. As well as reducing COD and TOC levels, activated carbon treatment is usually also able to give significant reductions in the colour and in the taste and odour of waters. It must be appreciated, however, that activated carbon adsorption does not provide a solution for all situations in which organic contamination creates problems. Ozone can break down complex organics such as pesticides either into inorganic end products or into simpler organic compounds which are more readily removed by biological oxidation or by activated carbon treatment.

Dissolved solids

Most conventional forms of water and wastewater treatment have little or no effect on the total dissolved solids content of water so that in a re-use situation the accumulation of dissolved solids may well limit the number of cycles that are possible. In many parts of the world, brackish groundwaters are found with TDS levels in excess of those acceptable in potable supplies and in arid areas these groundwaters or seawater may be the only available source of water. There has thus been considerable interest in the development of processes for the removal of excessive dissolved solids from such sources and these processes also have some application in the removal of dissolved solids from sewage and industrial effluents. The scale of the problem is somewhat different in that seawater has a TDS level of around 35 000 mg/l whereas the TDS levels of most effluents are around 1000 mg/l.

The distillation of seawater in evaporators has long been an accepted procedure for obtaining high-purity water although the finished product is not acceptable for drinking water until it has been aerated and chemically treated. The capital and operational costs for distillation are very high, so that the process is normally only used in situations where alternative sources of water are unavailable. Most modern distillation plants operate on the multi-stage flash process (Figure 19.1) which is based on the preheating of a pressurized salt-water stream, composed of seawater feed and return brine solution, with final heating taking place in a steam-fed heat exchanger. The heated salt water is then released to the first chamber where pressure is reduced, thus allowing part of the water to flash vaporize to steam which is condensed in a heat exchanger fed with the incoming salt solution. The remaining salt water passes to the next stage which operates at a slightly lower pressure so that further evaporation occurs. Most plants have thirty to forty stages with a temperature

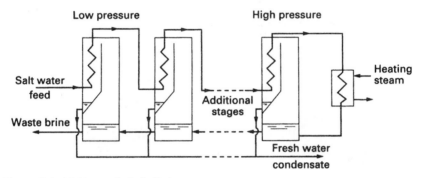

Figure 19.1 Multi-stage flash distillation.

range of ambient + 5°C to 110°C. Multiple-effect systems are returning to favour for smaller plants. Distillation costs depend upon the size of the plant and whether some of the steam can be used for electricity generation partly to offset production costs. The basic energy requirement for multi-stage flash distillation is about $200\,MJ/m^3$ of distillate but with dual-purpose distillation and power generation plants there is scope for varying the proportions of costs allocated to the two outputs. It is important to appreciate that as well as energy costs the production of freshwater from seawater by distillation involves other major costs for pretreatment, maintenance, labour and capital repayments.

In contrast to the complexity of flash-distillation installations, the use of solar stills has been studied in some parts of the world. These are low technology devices with a free source of energy but due to their low yield and the relatively high cost of the necessary glass structures the actual costs of water production are not greatly different from those of large flash-distillation plants.

Reverse osmosis (Figure 19.2) depends upon the phenomenon of osmosis in which certain types of membrane will permit the passage of freshwater whilst preventing or restricting the movement of soluble materials. Thus, if a semipermeable membrane is used as a barrier between a salt solution and freshwater the solvent (i.e. the water) will pass through the membrane to equalize the salt concentrations on either side. This movement occurs because of the osmotic pressure exerted by the dissolved salt. The process can be thought of as a form of hyper-filtration in which water molecules are small enough to pass through the pores in the membrane but larger molecules are unable to do so. The osmotic pressure is directly proportional to the concentration of the solution and to the absolute temperature and for seawater the osmotic pressure is about 24 bar (2.4 MPa). If a salt solution is subjected to a pressure greater than its osmotic pressure, water will pass through the membrane giving a desalted product and

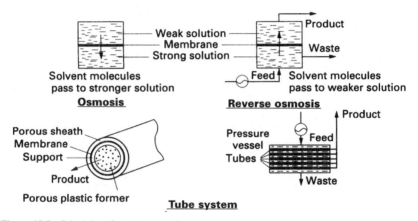

Figure 19.2   Principles of reverse osmosis.

leaving a concentrated brine. In practice, to obtain a significant yield of desalted seawater it is necessary to operate at pressures of 50–100 bar (5–10 MPa) and even then yields are only of the order of 0.5–2.5 m^3/m^2 day. Membranes were originally made from a cellulose acetate but most now have a polyamide base which allows higher flow rates and greater resistance to chemical degradation. Most commercial units utilize tubular systems to support the membranes at the high pressures necessary for economic operation. Modern membranes can give 99 per cent salt rejection so that a product of less than 500 mg/l TDS can be obtained from seawater in a single pass. Energy requirements for reverse osmosis are around 5–7 kW h/m^3 of product, but the high cost of the membranes, which have a limited life, means that the final cost of reverse osmosis-produced water is not far removed from that of flash distillation when used for seawater. A typical cost breakdown for a modern seawater reverse osmosis plant is shown in Table 19.6.

Table 19.6. Reverse osmosis costs (40 Ml/day plant)

Item	Cost (p/m³)
Chemicals, labour, maintenance	12
Membrane replacement (3 year life)	4.5
Electricity (5kW h/m³, 5.4 p/kW h)	27
Capital repayment (10%/year)	26.5
Total cost	70 p/m³

Reverse osmosis plants are becoming more popular for desalination than distillation because of their modular nature and reduced risks of corrosion problems. In order to protect the membranes and extend their life it is usually necessary to provide conventional water treatment before the supply is fed to the reverse osmosis units.

The ability of membranes to remove both soluble and fine suspended impurities has led to increasing interest in such processes to provide a satisfactory solution for a number of water and wastewater treatment problems. Reverse osmosis is normally considered as appropriate for the removal of soluble impurities in the molecular size range of about 5×10^{-3} to $5 \times 10^{-1}\,\mu m$. Ultrafiltration or nanofiltration is used to describe the removal of impurities in the size range of about 5×10^{-2} to $1\,\mu m$ with operating pressures of 1–10 bar (0.1–1.0 MPa). Microfiltration describes membranes with removals in the size range 10^{-1} to $5\,\mu m$ with operating pressures of 0.5–1 bar (0.05–0.10 MPa). Clearly there are overlaps between the different membrane processes in terms of the sizes of impurities removed and also in the types of membranes employed. Membrane manufacturers are able to produce systems which will reject dissolved substances above a specified molecular weight so that they can be custom made for a particular application. Nanofiltration membrane systems have considerable potential for removal of colloidal solids including microorganisms and soluble substances such as natural colour from both waters and wastewaters. Commercial units are now available with membranes to remove substances with molecular weights of greater than 200–400 Da when operating at pressures of 5–7 bar (0.5–0.7 MPa). Such membranes can reduce colour from 75°H to below 15°H as well as providing satisfactory disinfection although there is, of course, no residual effect, but this is not a major problem in small compact distribution systems.

The cross-flow arrangement as shown in Figure 19.3 allows the product water to escape at right angles to the direction of the feed stream flow and thus tends to reduce physical fouling of the membrane surface.

Removal performance is sometimes improved by the addition of a dynamic membrane formed on the inside of the fixed membrane by the controlled dosing of diatomaceous earth or calcium carbonate. These dynamic membranes can

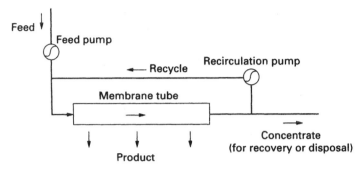

Figure 19.3 Cross-flow membrane system.

provide a higher rejection rate of impurities without reducing the flux rate to uneconomic levels and they are also claimed to prevent fouling problems due to biological growths. As well as removing large organic molecules, membrane processes will also remove microorganisms including viruses so that a high degree of disinfection can be achieved. By introducing powdered activated carbon or ion-exchange resins into cross-flow membrane systems it is possible to provide efficient contact systems with an effective solids/liquid removal stage. The properties of membranes have potential for providing solids/liquid separation in high-rate biological treatment systems where conventional clarification is not always satisfactory. Cleaning of flexible tube or sheet membrane systems is achieved by backwashing and/or physical pressure from roller devices. Membrane systems are normally constructed from modular units so that plants can easily be uprated by coupling in additional modules. By comparison with other solids/liquid separation processes, membrane systems are compact, although the overall cost of operation is likely to be similar to that of other processes.

For brackish groundwater with TDS of up to 3000 mg/l the electrodialysis process (Figure 19.4) is sometimes used although it is not suitable for handling seawater or supplies of similar salinity. Batteries of ion-selective membranes are placed in a cell so that when an electrical potential is applied, migration of ions occurs, giving alternate reduced-salinity and concentrated-salinity chambers. Pretreatment of the raw water is not so important as with the reverse osmosis process, although organic matter and sulphates can cause fouling of the membranes and iron and manganese should be removed to prevent their precipitation on the membrane surfaces.

On thermodynamic considerations, desalination by freezing compares favourably with distillation and, on an operational basis, low temperatures should give rise to fewer problems with corrosion and scaling. However, although a great deal of experimental work has been undertaken with freezing plants, many practical problems have arisen and the future of the process does not now seem very promising.

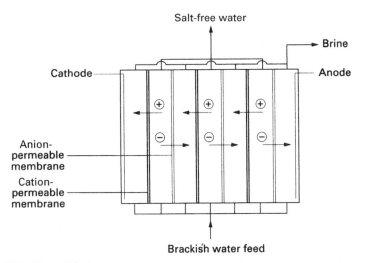

Figure 19.4 Electrodialysis.

All desalination processes produce a concentrated waste stream for which suitable disposal arrangements must be made.

Nitrogen compounds

The amount of ammonia present in sewage effluent can be reduced to low levels by biological nitrification either at the sewage works before discharge to the receiving water, which will prevent fish toxicity problems, or by pretreatment at the water works. In either case the process may be retarded by cold weather. The ammonia is of course oxidized to nitrate which is undesirable in other than small amounts in raw water supplies. Removal of ammonia by some other means may therefore prove to be more acceptable. Air stripping of ammonia can give good removals but tends to be very costly because of the large air flows required and the chemicals necessary to produce the high pH. Ion exchange using clinoptilolite to remove ammonia is possible although the exchange capacity of the material is relatively low. Considerable success has been achieved using ion-exchange resins to remove nitrate from groundwater supplies, although the available resins are not fully nitrate ion selective so that the cost and efficiency of this process depends upon the other ions present in the raw water.

High removals of nitrate produced by biological nitrification can be achieved by mixing recycled effluent with incoming settled sewage in a very low DO, or anoxic, environment. In these circumstances oxygen is removed from the nitrate by biological denitrification (Figure 19.5) so that most of the nitrate is converted to nitrogen which escapes to the atmosphere. Full-scale operation of activated-sludge plants with an anoxic zone at the inlet end has shown 80 per

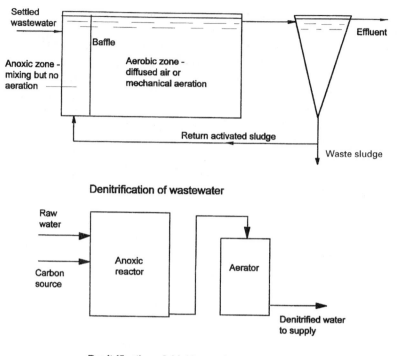

Figure 19.5 Biological denitrification processes.

cent or higher nitrate removals with a reduction in power costs due to the absence of aeration in the anoxic zone. Similar biological denitrification to remove nitrates from waters low in organic matter can be achieved in submerged bacteria beds or fluidized beds with the oxygen demand being provided by the addition of a cheap organic material such as methanol. However, such processes have proved to be costly and there are concerns about possible residuals of the organic feed stock in the finished water, particularly if the input nitrate concentration varies.

Phosphates

Phosphates in wastewater effluents are normally present as orthophosphates and can be fairly easily removed by chemical precipitation with aluminium or iron salts

$$Al^{3+} + PO_4^{3+} \rightarrow \underline{AlPO_4}$$
$$Fe^{3+} + PO_4^{3+} \rightarrow \underline{FePO_4}$$

The precipitation may be carried out at the primary sedimentation stage or in the final settling tanks. In some cases phosphates may be removed in an activated sludge plant by adding precipitants to the supernatant from anaerobically treated waste sludge. Effluent phosphate concentrations of 1–2 mg/l can be achieved reliably by precipitation although with the penalty of additional sludge production. To reduce phosphate concentrations below 1 mg/l an excess of reagent must be used which increases sludge production and chemical costs.

Biological removal of phosphorus can be achieved in systems which provide alternating anaerobic and aerobic reactors, analogous to the biological nitrogen removal process described earlier, and which usually remove nitrogen at the same time. A variety of proprietary process configurations have been developed and Figure 19.6 illustrates the Bardenpho and UCT systems originating from

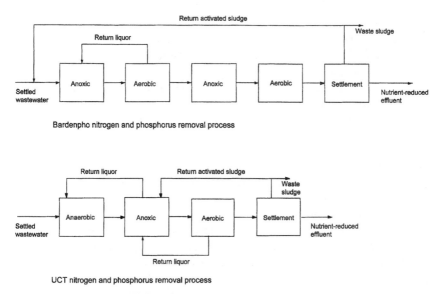

Figure 19.6   Biological nitrogen and phosphorus removal.

South Africa. These processes can increase the phosphorus removal in a conventional activated sludge plant of around 20 per cent to 90–95 per cent. Biological phosphate removal depends upon so called 'luxury uptake' of phosphorus by *Acinetobacter* which can assimilate volatile fatty acids (VFAs) produced under anaerobic conditions. The VFAs are then converted to polyhydroxyl butyrates (PHBs). Under aerobic conditions PHB is oxidized and the *Acinetobacter* can preferentially take up soluble phosphorus to be stored as polyphosphates. To ensure that the phosphorus remains with the biological solids it is essential to use aerobic stabilization for the sludge since under anaerobic digestion the phosphorus would be released back into solution and thus returned to the works inlet via the sludge supernatant. By suitable selection of operational conditions it is possible to remove both nitrogen and phosphorus or nitrogen and phosphorus independently. This can have economic significance where only one nutrient is limiting for algal growth in a site with eutrophication potential. It is important to note that biological nutrient removal processes are currently somewhat less stable and controllable than the chemical removal alternatives.

Microorganisms

Although conventional disinfection with chlorine compounds or other agents can produce satisfactory kills with very large concentrations of bacteria in reclaimed wastewaters, the DBPs formed are undesirable. The removal of most of the soluble organic contaminants, the precursors of DBPs, before disinfection will assist in achieving the required degree of disinfection without the formation of excessive amounts of undesirable DBPs but cannot entirely prevent their formation. Because of the resistance to chlorine compounds of *Cryptosporidium* and *Giardia* cysts, reclaimed waters must be viewed with some suspicion and the presence of viruses in such waters may also cause concern in some uses. Membrane treatment can provide a very high degree of removal of micro-organisms from reclaimed waters but the pressure drop across the membrane must be monitored continuously to warn of any pin-hole failures which could lead to microorganisms passing through into the product water.

19.4 Physico-chemical treatment of wastewater

In this discussion of tertiary treatment and water reclamation it has been assumed that wastewaters have initially been treated by conventional processes using physical and biological operations. Such processes are well developed and capable of providing reliable performance in most circumstances. Biological processes do, however, suffer from two possible disadvantages in that they cannot readily be switched on and off to meet intermittent loads and they are sensitive to toxic constituents in the wastewater. Partly because of these factors a considerable amount of research has been undertaken, mainly in the USA, to determine the performance capabilities of physico-chemical treatment plants using chemical coagulation and precipitation followed by filtration and adsorption. Such plants can produce effluents of around 10 mg/l BOD and 20 mg/l COD from relatively weak US wastewaters. In countries like the UK, where lower water consumptions give rise to stronger wastewaters, there is little evidence to suggest that physico-chemical plants could produce effluents of similar quality to those obtained from conventional plants and at the same cost. In the case of partial treatment requirements where discharge is made to coastal or estuarial waters where a relaxed effluent standard may be possible, physico-chemical treatment does, however, have some potential. Raw sewage dosed with a coagulant and then flocculated prior to sedimentation in an upward-flow sludge blanket settling tank can be transformed into an effluent with low SS and BOD. The effluent contains bacteria concentrations several orders of magnitude lower than in the raw sewage. The relatively small size occupied by this type of plant can be an advantage in resort areas, but it remains to be seen whether environmental legislation will permit the somewhat lower effluent quality than would be achieved by a conventional biological system.

References

Lauer, W. C., Rogers, S. E. and Ray, J. M. (1985). The current status of Denver's Potable Water Reuse Project. *J. Am. Wat. Wks Assn*, **77**(7), 52.

Metzler, D. F., Culp, R. L., Stoltenburg, H. A. *et al.* (1958). Emergency use of reclaimed water for potable supply at Chanute, Kan. *J. Am. Wat. Wks Assn*, **50**, 1021.

Pugh, N. J. (1945). Treatment of doubtful waters for public supplies. *Trans. Inst. Wat. Engnrs*, **50**, 80.

United States Environmental Protection Agency (1992). *Guidelines for Water Reuse*, EPA/625/R-92/004. Cincinnati: USEPA.

van Vuuren, L. R. J., Clayton, R. J. and van der Post, D. C. (1980). Current status of water reclamation at Windhoek. *J. Wat. Pollut. Control Fedn*, **52**, 661.

World Health Organization (1989). *Health Guidelines for the Use of Wastewater in Agriculture and Aquaculture*, Technical Report Series 778. Geneva: WHO.

Further reading

Argo, D. C. (1980). Cost of water reclamation by advanced wastewater treatment. *J. Wat. Pollut. Control Fedn*, **52**, 750.

Benneworth, N. E. and Morris, N. G. (1972). Removal of ammonia by air stripping. *Wat. Pollut. Control*, **71**, 485.

Bindoff, A. M., Treffry-Goatley, K., Fortmann, N. E. *et al.* (1988). The application of cross-flow microfiltration technology to the concentration of sewage works sludge streams. *J. Instn Wat. Envir. Managt*, **2**, 513.

Bouwer, H. (1993). From sewage farm to zero discharge. *Eur. Wat. Pollut. Control*, **3**(1), 9.

Burley, M. J. and Melbourne, J. D. (1980). Desalination. In *Developments in Water Treatment*, Vol. 2 (W. M. Lewis, ed.). Barking: Applied Science Publishers.

Buros, O. K. (1989). Desalting practices in the United States. *J. Am. Wat. Wks Assn*, **81**(11), 38.

Cooper, P., Day, M. and Thomas, V. (1994). Process options for phosphorus and nitrogen removal from wastewater, *J. Instn Wat. Envir. Managt*, **8**, 84.

Crook, J. (1985). Water reuse in California. *J. Am. Wat. Wks Assn*, **77**(7), 60.

Dean, R. B. and Lund, E. (1981). *Water Reuse*. London: Academic Press.

Dykes, G. M. and Conlon, W. M. (1989). Use of membrane technology in Florida. *J. Am. Wat. Wks Assn*, **81**(11), 43.

Gauntlett, R. B. (1980). Removal of nitrogen compounds. In *Developments in Water Treatment*, Vol. 2 (W. M. Lewis, ed.). Barking: Applied Science Publishers.

Harremoes, P. (1988). Nutrient removal for marine disposal. *J. Instn Wat. Envir. Managt*, **2**, 191.

Harremoes, P., Sinkjaer, O. and Hansen, J. L. (1992). Evaluation of methods of nitrogen and phosphorus control in sewage effluent. *J. Instn Wat. Envir. Managt*, **6**, 52.

Harrington, D. W. and Smith D. E. (1987). An evaluation of the Clariflow process for sewage treatment. *J. Instn Wat. Envir. Managt*, **1**, 325.

Hobbs, J. M. S. (1980). Sea water distillation in Jersey and its use to augment conventional water resources. *J. Instn Wat. Engnrs Scientists*, **34**, 115.

Laine, J.-M., Hagstrom, J. P., Clark, M. M. and Mallevialle, J. (1989). Effects of ultrafiltration membrane composition. *J. Am. Wat. Wks Assn*, **81**(11), 61.

Lauer, W. C., Rogers, S. E., La Chance, A. M. and Nealey, M. K. (1991). Process selection for potable reuse: Health effect studies. *J. Am. Wat. Wks Assn*, **83**(11), 52.

Mara, D. D. and Cairncross, S. (1989). *Guidelines for the Safe Use of Wastewater and Excreta in Agriculture and Aquaculture*. Geneva: WHO.

Mills, W. R. (1994). Groundwater recharge success. *Wat. Envir. Technol.*, **6**, 34.

Morin, O. J. (1994). Membrane plants in North America. *J. Amer. Wat. Wks Assn*, **86**(12), 42.

Polprasert, C. (1996). *Organic Waste Recycling* (2nd edn). Chichester: John Wiley.

Richard, Y. R. (1989). Operating experiences of full-scale biological and ion-exchange denitrification plants in France. *J. Instn Wat. Envir. Managt*, **3**, 154.

Rogalla, F., Ravarini, P., de Larminat, G. and Coutelle, J. (1990). Large-scale biological nitrate and ammonia removal. *J. Instn Wat. Envir. Managt*, **4**, 319.

Short, C. S. (1980). Removal of organic compounds. In *Developments in Water Treatment*, Vol. 2 (W. M. Lewis, ed.). Barking: Applied Science Publishers.

Shuval, H. I. (1987). Wastewater reuse for irrigation: evolution of health standards. *Wat. Qual. Bull.*, **12**, 79.

Tebbutt, T. H. Y. (1971). An investigation into tertiary treatment by rapid filtration. *Wat. Res.*, **5**, 81.

Thomas, C. and Slaughter, R. (1992). Phosphate reduction in sewage effluents: Some practical experiences. *J. Instn Wat. Envir. Managt*, **6**, 158.

Upton, J., Fergusson, A. and Savage, S. (1993). Denitrification of wastewater: operating experiences in the US and pilot plant studies in the UK. *J. Instn Wat. Envir. Managt*, **7**, 1.

Wade, N. and Callister, K. (1997). Desalination: State of the art. *J. C. Instn Wat. Envir. Managt*, **11**, 87.

Winfield, B. A. (1978). The performance and fouling of reverse-osmosis membranes operating on tertiary-treated sewage effluent. *Wat. Pollut. Control*, **77**, 457.

Yoo, R. S., Brown, D. R., Pardini, R. J. and Bentson, G. D. (1995). Microfiltration: A case study. *J. Amer. Wat. Wks Assn*, **87**(3), 38.

20

Water supply and sanitation in developing countries

In developed countries the public expect, and usually get, a high standard water-supply service and efficient collection, treatment and disposal of wastewaters. The techniques for pollution control are, in general, well developed and since populations are in low or zero-growth states demands on water resources are usually manageable. The picture is very different in developing countries where some 1.3 thousand million people are without safe water and more than 2 thousand million do not have adequate sanitation; this means that about 70 per cent of the population in these parts of the world lack basic facilities. The cost, in both monetary and manpower terms, of rectifying the situation will be large.

20.1 The current situation

As a result of the WHO's International Drinking Water Supply and Sanitation Decade (1981–90) large numbers of people in developing countries were provided with water and sanitation but rapid population growth has masked the improvements in many areas, as shown in Table 20.1.

In its monitoring of the decade, WHO concluded that a total of around US$135 thousand million had been invested in water supply and sanitation during the

Table 20.1. Water and sanitation in the International Drinking Water Supply and Sanitation Decade, 1981–90

	Millions of people without			
	Safe water supply		Adequate sanitation	
	1981	1990	1981	1990
Urban	213	243	292	377
Rural	1613	989	1442	1364
Total	1826	1232	1734	1741

period, 55 per cent on water supply and 45 per cent on sanitation. To achieve safe water and adequate sanitation for all by the year 2000 would require an annual investment of around US$50 billion which would be five times that achieved during the decade. It thus appears that the combined effects of shortages of financial resources, lack of trained personnel and the continuing population growth in several large developing countries will prevent the objectives of 'Safe Water for All 2000' from being achieved.

Although for urban areas throughout the world the ultimate aim may be to provide a developed country level of service, the provision of services at this level for the millions in rural areas of developing countries is unrealistic. Thus whilst the principles and processes discussed in this book are suitable for application in developed countries and as a target for all urban areas, more appropriate techniques must be adopted for rural communities in low-income areas. It is important to appreciate that to reduce the toll of water-related disease, improvements must be made in both water supply and sanitation although unfortunately sanitation is often neglected in favour of the more attractive water-supply activities. It is equally important to understand that the construction of sophisticated developed-country style water and wastewater treatment facilities, often favoured by donor governments and agencies, is of little value if the appropriate operation and maintenance back-up is not also provided. Many successful schemes have achieved their objectives by a combination of appropriate technologies and self-help by the communities served. This is a concept which has been undertaken with great success by the UK water industry charity WaterAid.

20.2 Sources of water

In the developed countries it is normal to provide at least some degree of treatment for water from any source, whereas for rural schemes in developing countries treatment will not be feasible in many circumstances. It is thus necessary to consider water sources in relation to what is likely to be the most important quality parameter, that of bacteriological quality.

Rainwater

With reasonably reliable rainfall the collection and storage of runoff from roofs can give a quite satisfactory source of water provided that the first flush of water from a storm, which is likely to be contaminated by bird droppings, etc., can be diverted away from the storage tank. With irregular rainfall, the size and cost of storage tanks may be large and unless the tanks are protected from contamination and the entry of mosquitoes, health problems can arise. Depending upon the intensity of the rainfall and the efficiency of the gutter and downpipe system, between 50 and 80 per cent of the rainfall may be collected.

Springs

Spring water is normally of good quality provided that it is derived from an aquifer and is not simply the discharge of a stream which has gone underground for a short distance. It is important to maintain this good quality by protecting the spring and its surroundings from contamination by humans and animals. A collecting tank should be constructed to cover the eye of the spring and prevent debris being washed into the supply.

Tube wells

Because of the natural purification, which removes suspended matter such as bacteria, groundwaters are usually of good bacteriological quality. Care must, however, be taken to ensure that sanitation practices, or the lack of them, do not cause groundwater contamination. Driven wells formed by well points in suitable ground conditions are relatively cheap although they often have a limited life due to corrosion of the tube and clogging of the perforations with soil particles. In sandy soils, tube wells made from plastic pipes may be rapidly driven by jetting. Bored wells can be produced by hand auger or by machine. Small diameter wells (40–100 mm) are normally fitted with simple hand pumps at ground level when the water table is sufficiently close to the surface and a great deal of work has been done by many organizations and manufacturers to produce a sturdy, reliable hand-pump design. For deeper water table sites, where a surface pump will have insufficient lift, the pump must be placed down the well, which usually requires a larger diameter bore and thus increases the cost. The head of a tube well should be provided with a suitable cap to prevent the entry of contaminated surfacewater.

Hand-dug wells

In many parts of the world hand-dug wells, 1–3 m diameter, are the traditional sources of water in rural areas. Depending upon the depth of the water table these wells may be as much as 30 m deep and during construction they can pose considerable hazards since the risk of collapse is often high. This risk can be greatly reduced by the use of precast concrete rings which sink as excavation proceeds and provide a permanent lining. In old wells the water quality often leaves much to be desired because of contamination due to the entry of surfacewater, spillage from containers and the deposition of debris. It is important that the site of the well is such as to avoid the entry of potentially contaminated groundwater and that a watertight lining extends for 3–6 m below the surface. The well head should have a headwall and drainage apron so that any surfacewater and/or spillage cannot gain entry to the well. These features are particularly important in areas where guinea-worm infections are endemic. Where possible, a pump should be used for water abstraction, thus permitting the use of a fixed cover on the well which will further reduce the risk of contamination of the water in the well.

Infiltration galleries

A porous collector system using open-jointed pipes in a gravel and sand filled excavation can be used to intercept high groundwater tables and give a further degree of filtration to the water. A similar arrangement can be usefully employed when abstracting water from rivers and lakes.

Surfacewater abstraction

The traditional developed-country sources of water in the form of rivers and lakes exist in many parts of the world, but in tropical countries the quality of surfacewaters is often poor so that for rural supplies it is advisable to use surfacewater only as a last resort.

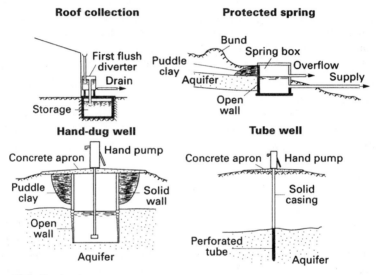

Figure 20.1 Rural water sources.

The basic characteristics of suitable rural water sources are shown in Figure 20.1. The undoubted attractions of groundwater supplies as regards bacteriological quality have resulted in many rural water schemes based on tube wells and it is vital to appreciate that it is only possible to abstract groundwater at a rate not exceeding the natural recharge. Disregard of this basic principle has led to falling groundwater tables and exhaustion of wells in a number of developing countries.

20.3 Water treatment

In the previous section emphasis was placed on the need to obtain water from sources likely to be free from harmful levels of contamination. This is particularly important for rural water supplies since treatment processes greatly increase the cost of providing water and unless reliable operation and maintenance skills are available the treatment is soon likely to fail. There is no such thing as a maintenance-free treatment process and for developing-country situations, water treatment should not be adopted unless its use is unavoidable. In this context it is not realistic with small rural supplies to adopt the normal requirement of no *E. coli* in 100 ml of water since this would imply the need for treatment for many sources. There is considerable evidence that waters with up to 1000 *E. coli* per 100 ml can be supplied to rural communities with little or no health hazard to the resident population. Indeed, the provision of such a poor-quality (by developed-country standards) water could well bring considerable improvements in health in the many areas where water-washed diseases are the major source of ill health. When considering alternative sources of water it is worth remembering that a good-quality source some distance away from the community could possibly be conveyed to the community at a lower long-term cost than that for treating a nearer but poorer-quality source and the former solution is certainly likely to be more reliable.

If there is no alternative to the provision of treatment every effort must be made to keep the treatment as simple as possible to try to ensure low cost, ease of construction, reliability in operation and to enable operation and maintenance to be satisfactorily undertaken by local labour. Failure to satisfy these basic aims will almost certainly produce many problems and will often lead to the abandonment of the scheme with reversion to the traditional unimproved sources.

Storage

Storage can provide a useful measure of purification for most surfacewaters although it cannot be relied upon to produce much removal of turbidity. The disinfecting action of sunlight normally gives fairly rapid reductions in the numbers of faecal bacteria. To obtain the maximum benefit from storage it is important to ensure that short-circuiting in the basin is prevented by suitable baffles. A disadvantage of storage in hot climates is the considerable evaporation losses which can occur. The design of storage facilities should be such as to prevent the formation of shallow areas at the edges which could provide mosquito-breeding sites. If the settlement provided by storage does not give sufficient removal of suspended matter the choice for further treatment involves consideration of chemical coagulation and/or filtration techniques.

Coagulation

The use of chemicals for coagulation brings a further level of complexity to the treatment process and should only be adopted if the necessary supplies and skills are available locally. The use of natural coagulants like *Moringa oleifera* as described in Chapter 12 can make coagulation feasible in situations where conventional coagulants are unaffordable or unavailable. Chemical coagulation will only be successful if the appropriate dose can be determined and then applied to the water in such a manner as to ensure adequate mixing and flocculation. A simple form of chemical feeder is one based on hydraulic control of a solution such as the Marriotte vessel which provides a constant rate of discharge regardless of the level in the storage container. The coagulant must be added at a point of turbulence such as a weir or in a baffled channel and flocculation is best achieved in a baffled basin connected to a settling basin. In practice it is difficult to prevent floc carry-over from the settling basin so that the output water quality may at times have fairly high turbidity levels. A more satisfactory water may therefore arise by omitting the coagulation stage and proceeding directly with filtration.

Filtration

Although some simplified types of rapid sand filter are available, the slow sand filter is likely to be the most satisfactory form of treatment process for many developing-country installations, certainly in rural areas. Slow filtration is able to provide high removals of many physical, chemical and bacteriological contaminants from water with the advantages of simplicity in construction and use. No chemicals are required and no sludge is produced. Cleaning by removal of the top surface at intervals of a month or more is labour intensive, but this is not normally a problem in developing countries. The relatively large area requirements for a slow filter are unlikely to cause problems for small supplies and the cost of construction can be reduced by using locally available sand or substitute materials such as rice husks. It is usually possible to operate slow filters at rates of about $5 \, m^3/m^2$ day with peak raw water turbidities of up to 30 NTU whilst producing a filtrate of <1 NTU.

Horizontal-flow gravel bed filters can be very effective in providing pretreatment of waters with a turbidity >30 NTU to allow slow sand filters to be used to provide a high quality product water with reasonably long filter runs.

Disinfection

If disinfection is required, the previously stated problems related to chemical dosing must again be faced. Chlorine is the only practicable disinfectant, but the availability of the gaseous form is likely to be restricted and in any event the hazards of handling chlorine gas make it undesirable for rural supplies. A more suitable source of chlorine for such installations is bleaching powder which is about 30 per cent available chlorine and is easy to handle, although it loses its strength when exposed to the atmosphere and to light. High-test hypochlorite (HTH) in granular or tablet form has a higher available chlorine content (70 per cent) and is stable in storage but more costly. Sodium hypochlorite solution is another possible source of chlorine. Dosage should be by means of the same type of simple hydraulic feeder discussed for coagulants. For very small supplies, simple pot chlorinators using a pot or jar with a few tiny holes and filled with a mixture of bleaching powder and sand can give a chlorine residual for about two weeks before recharging is necessary. The attainment of the correct dose of chlorine is important since too low a dose will give a false sense of security and too high a dose will give a chlorine taste to the water which will probably result in rejection of the supply by the consumers in favour of traditional but less safe sources.

20.4 Sanitation

The relationships between a considerable number of water-related diseases and the presence in the environment of excreta from people suffering from these diseases are well established. It could indeed be argued that the sanitary disposal of human excreta is more important in a health context than the provision of a safe water supply. Even in the presence of good-quality water, direct faecal–oral contact can maintain high levels of incidence of diseases such as typhoid and cholera. It would therefore seem important to make every effort to prevent faecal contamination of water sources as a primary objective since the treatment of an already polluted water can be costly and, particularly on a small scale, is unlikely to have high reliability.

Excretion is inevitably a highly personal process and as such is largely governed by the sociological patterns in a particular community. A vital first step in any sanitation programme is therefore to gain a full understanding of current excretion practices and of the likely acceptability of possible alternatives. It is generally true in rural areas that excreta disposal is far more complex socially than it is technically. A purely engineering solution may well be quite unsatisfactory because its sociological implications have not been examined. Improvements in public health do not necessarily follow the installation of a sanitation system since unless the new facilities are correctly used and given the appropriate level of maintenance, little benefit will arise.

When considering the various types of sanitation systems a basic differentation can be made between dry systems, which essentially handle only faeces, possibly with some urine, and wet systems which handle faeces, urine and sullage (the liquid wastes from cooking, washing and other household operations). A simple classification of sanitation methods, the more important of which are shown schematically in Figure 20.2, would be

- dry, on-site treatment and disposal – trench and pit latrines, composting latrines

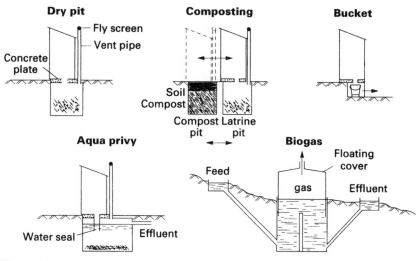

Figure 20.2 Simple sanitation systems.

- dry, off-site treatment and disposal – bucket or vault latrine with collection service and central treatment facility
- wet, on-site treatment and disposal – wet pit, aqua privy, septic tank, biogas, land disposal
- wet, off-site treatment and disposal – conventional or modified sewerage and central treatment facility.

Pit latrines

These provide the simplest form of latrine and are widely used because of their simplicity, low cost and ease of construction in suitable ground conditions. The usual form has a pit about 1 m square and 3–4 m deep. A volume of 0.06–0.1 m³/person year is often used to estimate the life of a pit and when it is about two-thirds full it is filled in with soil and the superstructure transferred to a new pit.

If a single pit is to be emptied for further use, off-setting it by using an inclined chute enables more convenient operation. The provision of a vent pipe from the pit to a height above the top of the superstructure, the VIP latrine, will do much to reduce smells and control insects which otherwise may discourage use of the latrine. A variant of the pit latrine uses a bored hole 200–400 mm in diameter and perhaps 6 m deep. The life of such a unit is likely to be less than that of a pit and fouling of the sides of the bore often produces odour problems. Unlined pits for excreta disposal must only be used in situations where there is no danger of groundwater contamination. They should always be placed downhill of any water source and not within 30 m of a well used for water supply.

Composting latrines

In some developing countries the fertilizer value of human excreta may be a significant factor in crop production so that sanitation methods which permit excreta re-use for such purposes are appropriate. It is highly desirable that pathogenic organisms in the excreta are destroyed before the material is applied to land and crops since otherwise the potential for the spread of disease is considerable. The required destruction of pathogens can be achieved by composting excreta, vegetable scraps, grass cuttings etc., usually in a batch process with relatively long retention time or possibly in a continuous composting unit with a retention time of a few months. For effective composting a C:N ratio of between 20 and 30 to 1 is necessary and the moisture content must be within the range 40–60 per cent. Batch composters are typified by the double-vault units widely used in South East Asia where the latrine is built on top of two bins which serve in turn as receptacles for faeces, cleaning paper and wood ash. Ash amounting to about one-third of the weight of faeces is normally sufficient to prevent odours. Urine is collected separately and used directly on the land. When a bin is about two-thirds full the contents are levelled, covered with earth and the bin sealed. Defaecation is transferred to the second bin and the first bin left for a period of up to twelve months before being emptied. A bin capacity of $0.4\,m^3$ per person is often recommended based on a one-year cycle of operation.

Although a number of continuous composting systems have been produced in the developed countries their performance in developing countries has not been very satisfactory. The proper operation of a composting toilet needs careful attention, particularly in regard to the moisture content, and they should not be used in preference to a pit latrine unless the requisite degree of supervision can be provided.

Bucket and vault latrines

The removal of excreta from latrines in a variety of containers is one of the oldest forms of sanitation and is still widely used in many parts of the world. The traditional bucket latrine utilizes a squatting plate or pedestal set above a metal bucket in a chamber with a door to the outside of the house. The bucket is emptied, usually at night (hence the term nightsoil), into a larger container carried by hand or on a cart to a collection depot or disposal area. The system as normally operated has little to commend it from a hygienic aspect. Spillages during handling occur frequently and the buckets are rarely cleaned so that fly nuisances are common. With improvements such as lids for the buckets during handling, washing and disinfection of the buckets and well-designed and maintained latrines the system could be made more acceptable.

A development of the bucket system uses water-tight vaults below the latrine which are emptied at intervals of about two weeks using a suction tanker which may be hand operated or mechanized. With proper design and maintenance such a system can be quite satisfactory but the cost and complexity of the removal system makes its suitability for developing countries rather limited, although it is widely used in Japan.

Ultimate disposal of the collected waste is usually achieved by burial in shallow trenches, often hand dug to a size of $4 \times 1 \times 0.5$ m. These are filled with nightsoil to a depth of about 0.3 m and back filled with soil. The area can be re-used after a period of at least twelve months. Unfortunately, in some areas the nightsoil is simply dumped on waste ground with no attempt at burial, thus producing highly undesirable environmental conditions.

Wet pit latrines

In a number of developing countries the pour-flush water-seal latrine is popular. Water usage to maintain the seal is 1–3 l per occasion so that the contents of the pit become semi-liquid. Anaerobic digestion of the contents occurs, thus reducing their volume to some extent so that a design volume of 0.04–0.06 m^3/person year should be suitable. The pit is usually placed a short distance away from the latrine and connected to it by a short length of steeply sloping 100 mm pipe. The pit is usually lined with open-jointed brickwork to prevent collapse but to allow percolation of the liquid contents into the surrounding ground. The water seal means that fly and odour nuisances are prevented so that such a latrine is suitable for indoor installation. Success of a wet pit system depends upon a relatively low level of water usage and the presence of suitable ground conditions to allow escape of the liquid without causing groundwater pollution.

Aqua privies

An aqua privy consists of a small tank situated below a latrine and discharging an anaerobically treated liquid effluent which must be suitably disposed of to complete the system. The latrine plate or pan must have a submerged down pipe or water trap to maintain anaerobic conditions in the tank, prevent the escape of odours and stop the entry and egress of insects. A typical tank capacity would be 0.12 m^3/person with a sludge accumulation of about 0.04 m^3/person year. The volume of effluent discharged is likely to be about 6 l/person day. A major operating problem which has been found with many aqua privy installations is that insufficient water is added to maintain the water seal so that the latrine becomes unattractive to the user. Unless evaporation and leakage losses can be made up the liquid level in the tank falls, breaking the seal and giving odour and insect problems. In some installations sullage water is piped to the tank to provide the necessary make-up volume; in this case an additional 0.5 m^3 capacity should be provided to allow for the sullage.

Septic tanks

As indicated above, aqua privies are basically simplified septic tanks, which are widely used throughout the world in many rural areas of developed countries. As anaerobic units they provide the removal of much of the suspended matter from sewage, the deposited solids then digesting with consequent release of some of the organics in soluble form. A typical septic tank will remove about 45 per cent of the applied BOD and about 80 per cent of the incoming SS. The effluent will contain large numbers of bacteria and concentrations of 10^6 *E. coli*/100 ml are not uncommon in the discharge from a septic tank. Solids accumulations of around 0.05 m^3/person year are often assumed and the sludge must be removed at intervals of 1–2 years. The sludge is very strong with BOD and SS levels of between 10 000 and 50 000 mg/l. Typical design criteria are a minimum liquid retention time of three days and for populations of between 4 and 300, UK practice is to provide a volume in m^3 of ($0.18 \times$ population + 2). Again it is important to provide suitable inlet and outlet arrangements to ensure a water seal and prevent the discharge of the scum layer which usually exists on the top of the liquid.

Biogas

The utilization of methane produced by anaerobic digestion of wastewater sludges is well established in developed countries, normally accompanied by a great deal of complex and expensive plant. However, any container holding putrescible organic matter will develop anaerobic conditions with the consequent production of methane. Thus an individual dwelling aqua privy or septic tank will develop a small amount of methane, but the volume is insufficient for any significant use. In rural areas in developing countries the collection of solid wastes from man and animals in a simple container can produce sufficient gas for domestic light and cooking requirements and the digested sludge has a high nitrogen content, making it a useful fertilizer. Many biogas installations constructed from simple materials are in use in the Far East. Most of these plants are quite small, 1–5 m^3 capacity, and are operated by individual farmers. The daily gas requirement for cooking is about 0.2 l/person day and this requirement can probably be satisfied by digestion of human excreta plus the excreta from a single cow or similar beast. The flammable nature of methane gas means that even small units must be carefully sited and operated to avoid the risk of serious explosions. Design loadings for biogas units are around 2.5 kg VS/m^3 day with a nominal retention time of 20 days or more.

20.5 Effluent disposal

Depending upon the choice of sanitation technology, appropriate provision must be made for disposal of any liquid effluents.

On-site treatment and disposal

For all wet sanitation systems safe disposal of the liquid effluent is an inherent part of the system. The effluent from such systems is likely to be relatively low in SS but will probably have a high organic content and large numbers of microorganisms so that indiscriminate release to the environment would create health hazards. In many circumstances the most satisfactory method of disposal is into the ground via seepage pits, drainfields or evapotranspiration beds (Figure 20.3). The applicability of the various methods is related to soil permeability, groundwater levels and proximity to buildings. If sub-surface disposal is not possible, further treatment of the effluent in a facultative oxidation pond or a simple bacteria bed may be feasible for a community. Septic tanks, aqua privies and soakaways should normally be sited at least 30 m away from wells and boreholes, on the downstream side of the flow, and similar distances should be used in relation to sites near surfacewater sources.

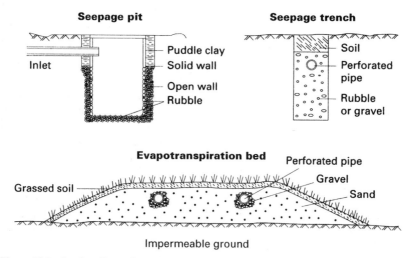

Figure 20.3 On-site effluent disposal systems.

A soakaway may be in the form of a seepage pit, with porous construction, in sites where the soil is highly permeable. The pit should be of similar size to the aqua privy or septic tank which it serves. In most soils, drainfields provide the most satisfactory form of sub-surface disposal. The system comprises seepage trenches containing open-joined or perforated pipes surrounded by stone fill and topped by an earth backfill. Most of the percolation occurs through the sides of the trench and a reasonable loading for many soils is $10 \, l/m^2$ day, based on sidewall area.

In areas where the ground is impermeable an artificial soakage area or evapotranspiration bed can be provided in a suitable depression, or above the general ground level in the form of a mound. The bed or mound is made of coarse sand and gravel surrounding the pipes and topped with a layer of soil supporting a fast-growing grass. Liquid losses from such an area will probably be about 80 per cent of the evaporation from a free water surface in the same locality.

Whilst sub-surface disposal methods can be quite satisfactory in low population-density areas the possible hazard to groundwater quality must be recognized and in urban areas it is unlikely that sufficient land area will be available for the methods to operate correctly.

Off-site treatment and disposal

In urban areas it may be necessary to install sewerage systems to collect all liquid wastes and convey them to a treatment facility. Unless a reliable water supply is available a conventional sewerage system will have many problems due to solids deposition in low flows, hydrogen sulphide production and consequent corrosion effects being particularly troublesome. The capital cost of a conventional sewerage system is very high and the disruption caused by its construction in a congested urban area must also be considered. In some areas a modified sewerage system has been adopted to collect the effluent from individual septic tanks and aqua privies. In this situation the flows are likely to be relatively low and since the bulk of the solids have been removed, small diameter pipes laid at shallow gradients will suffice, greatly reducing construction problems and costs. It must be appreciated that such a modified sewerage system will only work satisfactorily if the individual tanks are regularly desludged and gross solids are prevented from entering the system.

With a sewerage system it will be necessary to provide some form of central treatment facility to ensure that the effluent can be discharged without causing significant environmental damage. In most developing country situations the best form of wastewater treatment will probably be provided by facultative oxidation ponds which are simple to construct and operate. They do, however, require considerable amounts of land so that in urban areas this may be a problem. Care must be taken to prevent shallow water and vegetation at the edges providing conditions attractive for mosquitoes and other insects. The large algal growth in such ponds usually means that the SS in the effluent are relatively high due to escaping algae. In some areas the possibility of protein recovery from algal matter may be worth consideration. If conditions are unsuitable for oxidation ponds, bacteria beds or activated sludge units may be necessary, but these relatively complex installations should only be adopted if the appropriate level of operational and maintenance skills can be assured and the necessary financial support is also available.

Further reading

Bailey, R. A. (ed.) (1996). *Water and Environmental Management in Developing Countries.* London: CIWEM.

Cairncross, S. and Feachem, R. G. (1983). *Environmental Health Engineering in the Tropics.* Chichester: John Wiley.

Cairncross, S. and Ouano, E. A. R. (1991). *Surface Water Drainage for Low Income Communities.* Geneva: World Health Organization.

Carter, R., Tyrell, S. F. and Howsham, P. (1993). Lessons learned from the UN water decade. *J. Instn Wat. Envir. Managt,* **7**, 646.

Diamant, B. Z. (1979). The role of environmental engineering in the preventive control of waterborne diseases in developing countries. *Roy. Soc. Hlth J.,* **99**, 120.

Franceys, R., Pickford, J. and Reed, R. (1992). *A Guide to the Development of On-site Sanitation.* Geneva: WHO.

Franklin, R. (1983). *Waterworks Management in Developing Countries.* Morecambe: Franklin Associates.

Glennie, C. (1983). *Village Water Supply in the Decade.* Chichester: John Wiley.

Hutton, L. G. (1983). *Field Testing of Water in Developing Countries.* Medmenham: WRC.

International Reference Centre for Community Water Supply and Sanitation (1981). *Small Community Water Supplies.* The Hague: IRC.

Kalbermatten, J. M. (1981). Appropriate technology for water supply and sanitation: Build for today, plan for tomorrow. *Pub. Hlth Engnr,* **9**, 69.

Lee, M. and Bastemeijer, T. (1990). *Drinking Water Source Protection.* The Hague: IRC.

Mara, D. D. (1976). *Sewage Treatment in Hot Climates.* Chichester: John Wiley.

Morgan, P. (1990). *Rural Water Supply and Sanitation.* London: Macmillan.

Pacey, A. (ed.) (1978). *Sanitation in Developing Countries.* Chichester: John Wiley.

Reed, R.A. (1995). Sustainable Sewerage: Guidelines for Community Schemes. London: Intermediate Technology Publications.

Schulz, C. R. and Okun, D. A. (1984). *Surface Water Treatment for Communities in Developing Countries.* New York: Wiley. (Reprinted 1992, London: Intermediate Technology Publications.)

Various authors (1980–82). *Appropriate Technology for Water Supply and Sanitation,* Vols 1–12. Washington DC: World Bank.

Vigneswaran, S. and Visvanthan, C. (1995). *Water Treatment Processes: Simple Options.* Boca Raton: CRC.

Wegelin, M. (1996). *Surface Water Treatment by Roughing Filters: A Design, Construction and Operation Manual.* London: Intermediate Technology Publications.

White, G. F., Bradley, D. J. and White, A. U. (1972). *Drawers of Water.* Chicago: University of Chicago Press.

Wolman, A. (1978). Sanitation in developing countries. *Pub. Hlth Engnr,* **6**, 32.

Index